Additive Manufacturing In 3D Polymers

Usha Bano

CHAPTER – 1

Introduction

Additive manufacturing (AM), also known as 3D printing, is a rapidly evolving technology that allows the creation of complex geometries with improved accuracy and reduced waste compared to traditional manufacturing processes. It is a layer-by-layer production process that builds a three-dimensional object from a digital model by adding material incrementally.

The history of AM dates back to the 1980s, when the first patent was filed for stereolithography, which uses a laser to solidify layers of resin into a 3D shape. Since then, numerous AM technologies have been developed, including fused deposition modelling (FDM), selective laser sintering (SLS), and electron beam melting (EBM), among others [1,2].

AM has the potential to revolutionize the manufacturing industry by enabling the production of complex geometries that are difficult or impossible to manufacture using traditional methods, such as subtractive manufacturing. It also offers the ability to produce parts on-demand, reducing lead times and enabling greater customization. One of the most significant advantages of AM is the ability to produce parts with improved material properties, particularly in the case of metals and alloys. For example, AM can produce materials with improved strength (both yield strength (YS) and Ultimate tensile strength (UTS)), and corrosion resistance compared to conventionally manufactured materials. The applications of AM are wide-ranging, including aerospace, biomedical engineering, automotive, and consumer goods, among others. As the technology continues to advance, it is expected to have an even greater impact on the manufacturing industry [3,4].

While the phrase "3D Printing" is often used as a substitute for all AM processes, there are actually numerous distinct layer-based methods that vary based on the materials used and the technology of the machines. Consequently, in 2010, the American Society for Testing and Materials (ASTM) established the "ASTM F42 – Additive Manufacturing" group, which created a set of standards that organizes the range of AM processes into the following seven categories (refer to Standard Terminology for Additive Manufacturing Technologies, 2012).

VAT Photo-polymerisation: This technique utilizes a container of liquid photopolymer resin, which is solidified one layer at a time with a light source, such as a laser or UV light. Stereolithography (SLA) and digital light processing (DLP) are the most prevalent forms of VAT Photo-polymerisation.

Material Jetting: This method employs inkjet printing technology to deliver droplets of material onto a build surface, which are subsequently cured with UV light. The process is similar to traditional paper printing but occurs in three dimensions. Technologies like PolyJet and MultiJet Printing exemplify material jetting.

Binder Jetting: This approach entails spreading a thin layer of powder material over a build platform and using a liquid binder to selectively adhere the powder particles together. This process is repeated layer by layer until the desired shape is achieved. Examples of binder jetting techniques include Selective laser sintering (SLS) and powder bed inkjet three-dimensional printing (3DP).

Material Extrusion: This technique involves melting a thermoplastic filament and pushing it through a nozzle to build a three-dimensional object layer by layer. Notable examples of material extrusion methods include fused deposition modelling (FDM) and fused filament fabrication (FFF).

Powder Bed Fusion: In this technique, a laser or electron beam selectively fuses a layer of powdered material to form a solid object. Common processes within powder bed fusion are selective laser melting (SLM) and electron beam melting (EBM).

Sheet Lamination: This method involves layering sheets of material (such as paper or metal) and cutting them into the desired shapes with a laser or knife. The sheets are then bonded or fused together through adhesion or heat. Examples include laminated object manufacturing (LOM) and ultrasonic additive manufacturing (UAM).

Directed Energy Deposition: This technique focuses on the melting and application of material onto a substrate using either a laser or an electron beam. The material is usually in filament or powder form, and this process is often utilized for repairing or enhancing existing components. Examples of directed energy deposition methods include laser-engineered net shaping (LENS) and direct energy deposition (DED).

Every AM process comes with its unique set of benefits and drawbacks (as detailed in Table 1.1). Choosing a particular process depends on various factors, including material characteristics, desired accuracy and resolution, production rate, and cost considerations. The following sections of this thesis focus on Powder Bed Fusion, particularly in relation to the additive manufacturing of IN-718.

The advantages of AM technology, which is particularly effective in producing customized and complex shapes, internal structures, heterogeneous materials, and products with anisotropic properties, which can be costly or even impossible to achieve using traditional manufacturing techniques [5]. It significantly reduces dependence on jigs, fixtures, and tooling, while providing a build setup that does not rely on specific parts, thus accelerating product development and optimizing the manufacturing process [3]. AM allows for the simultaneous creation of multiple designs and assemblies within a single build cycle [6,7]. AM offers substantial opportunities to improve material efficiency and reduce life cycle impacts and carbon emissions [8,9]. When it comes to the constraints of AM technology, the quality of the products tends to be lower than that produced by conventional subtractive manufacturing methods [3,10,11]. AM requires post-processing steps to achieve the necessary tolerances. The materials used in this process are costly, and the AM processing duration is often lengthy [12]. As a result, production expenses are greater when compared to traditional manufacturing methods [13]. Also, the dimensions of the products that can be printed are limited by the size of the building platform.

The iterative lifecycle stages for additively manufactured aerospace components can be generally divided into four phases: (i) design and pre-processing, (ii) manufacturing, (iii) post-processing, and (iv) in-service evaluation. Each of these phases, along with associated sub-steps, is essential for process selection and has a significant effect on the final performance of the part (refer to Figure 1.1). The design phase includes standard aerospace design methodologies such as computer-aided design (CAD) and comprehensive discipline analyses, as well as specific design-for-AM verification processes like model validations and part assessments to ensure that the exported model meets the original design requirements. After the design phase, the pre-processing stage focuses on establishing the build layout, generating support structures, slicing the model, and configuring build parameters, which encompass tool-paths, scanning strategies, and energy density settings. The design and pre-processing stages can be quite iterative, often necessitating multiple cycles before advancing to the manufacturing phase.

Table 1.1: Comparison of 7 AM processes [2,3]

AM Techniques	Advantages	Disadvantages	Applications
VAT Photo-polymerisation	• High accuracy and resolution • Good surface finish • No support structures needed	• Limited material and size capability • Slow production & Post-processing required to remove excess resin	• Jewellery & Dental applications
Material Jetting	• Multi-material capability • Good surface finish	• Limited size capability • Expensive equipment and materials	• Prototyping • Complex assemblies
Binder Jetting	• Fast production speed • Large build volume	• Poor accuracy and resolution • Limited material options • Porous parts • Requires post-processing	• Prototyping • Sand casting moulds • Architecture models
Material Extrusion	• Wide range of material • Large build volume	• Limited accuracy and resolution • Requires support structures	• Prototyping • Tooling • Low-volume production
Powder Bed Fusion	• High accuracy • Multi-material capability • High strength and density • Can produce complex geometries	• Expensive equipment and materials • Requires post-processing to remove excess powder	• Aerospace • Medical implants
Sheet Lamination	• Low cost • Large build volume • Multi-material capability	• Poor accuracy and resolution • Poor surface finish	• Prototyping • Architecture models
Directed Energy Deposition	• High accuracy and resolution • Can produce large parts • Multi-material capability	• Expensive equipment and materials • Poor surface finish & Requires post-processing	• Repair and maintenance of high-value components • Tooling

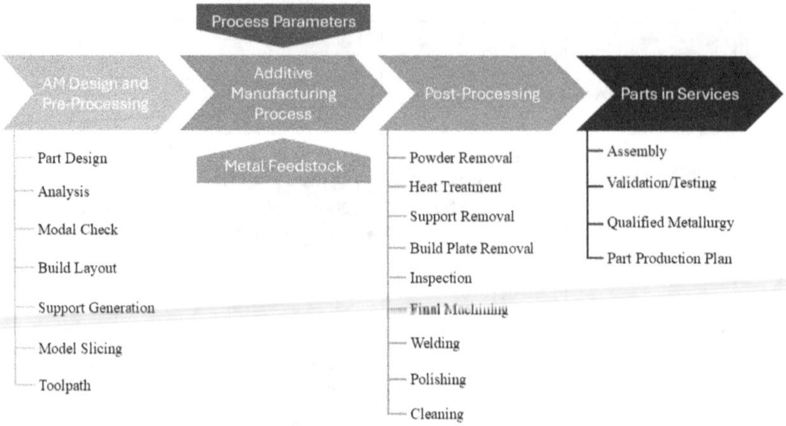

Figure 1.1: Iterative lifecycle stages for additively manufactured aerospace components [14]

The aerospace sector employs a diverse array of metal AM techniques, each known by specific terms and classifications. Factors like feedstock type, the physical state during fusion, material distribution techniques, and fundamental technological principles, including the energy source, are outlined in [15]. These AM techniques encompass powder bed fusion (PBF), DED, and solid-state methods such as cold spray (CS), additive friction stir deposition (AFS-D), and metal sheet lamination. These are employed in repairs, remanufacturing, coating applications, and the creation of freeform components. The metallic parts consist of pure metals, alloys, metal-matrix composites, and multi-alloy configurations. Figure 1.2 presents a series of close-up images highlighting the different metal AM processes and indicates the direction of deposition and build.

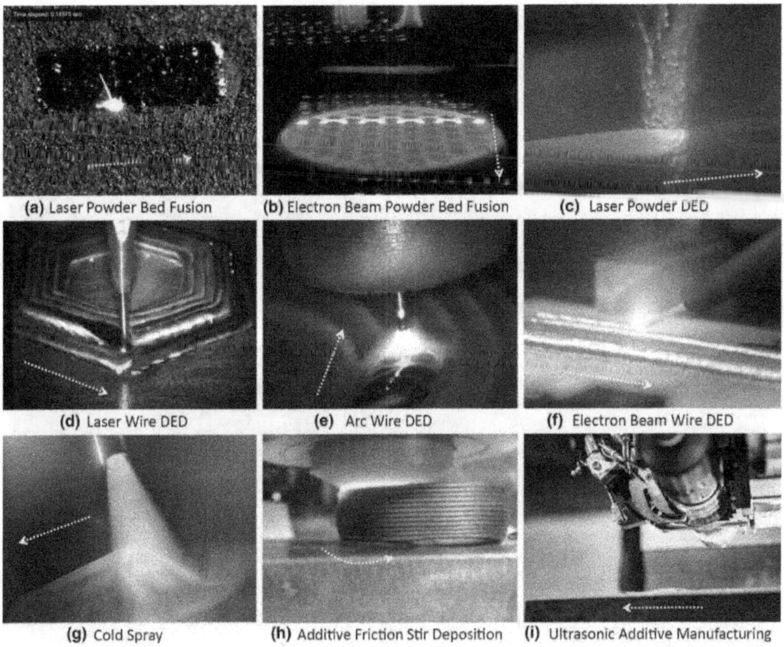

Figure 1.2: Images of metal AM processes along with deposition/build direction [14]

1.1 Laser Powder Bed Fusion (LPBF)

Laser powder bed fusion (LPBF) employs a powder bed and a heat source to produce metal components, providing significant design flexibility without the necessity for tooling. LPBF is particularly advantageous for high-cost materials, such as superalloys (IN718, IN625), titanium alloys (Ti-6Al-4V), and refractory materials (C103), as producing final components typically requires extensive machining with sophisticated CNC (computer numerical control) machines, which can be both expensive and time-consuming [16–18]. Enhancing the design of parts or assemblies through 'design for additive manufacturing' can lead to lighter printed components, minimize or eliminate weld joints, and improve overall cost-effectiveness, time efficiency, inspection processes, and reliability. AM is increasingly appealing to various contemporary manufacturing industries, including aerospace, defence, energy, and automotive sectors, where traditional subtractive methods struggle to fabricate materials like nickel-based superalloys [19]. To promote their broader application, the production of nickel-based superalloys using PBF has attracted significant interest.

This 'bottom-up' approach, which relies on laser melting metal powders layer by layer, often results in microscopic defects such as porosity, Lack of Fusion (LOF), and cracks. To address these issues, it is crucial to maximize relative density close to 100% [20,21]. Consequently, it is vital to methodically optimize the process variables that influence the quality of the parts being built, aiming for maximum densification and enhanced mechanical properties in components suitable for static, dynamic, and shock loading conditions at both ambient and extreme temperatures. Thus, understanding how process variables affect defect formation and how these defects impact material properties is essential [13,22–24]. Additionally, various metal AM processes are highlighted, illustrating the direction of deposition/build.

The LPBF process consists of several stages. First, a layer of metal powder is spread evenly across a build platform. The laser beam is then directed at specific locations within the powder layer, causing it to melt and solidify. The platform is lowered, and the process is repeated for the next layer. This cycle continues until the part is complete. The process is carefully controlled to ensure that the part is produced with the required precision, surface finish, and strength.

In the LPBF process, a laser beam is employed to selectively melt a thin layer of metal powder, shaping it according to the cross-section of the intended final component [25,26]. Once a layer is fused, the build platform is lowered, a fresh layer of powder is distributed, and this cycle repeats until the entire part is fabricated. Recently, Caiazzo et al. [27] investigated the significance of VED (volumetric energy density) in the context of IN718 within LPBF, they also detailed the technological and mechanical characteristics of AM components, including surface roughness, hardness, pore size and distribution, as well as fractional density, all in terms of VED.

LPBF systems incorporate numerous process variables, with the most frequently addressed being laser power (LP in W), scanning speed (SS in mm/s), powder bed layer thickness (PBLT in mm), and the distances between successive laser passes and hatch distance (HD in mm). The aim is to establish optimal process parameters that achieve dense builds, minimize defects, reduce surface roughness, enhance build speed, and ensure that produced parts meet acceptable material specifications [28]. Given the various design options for a part, determining a universally optimized set of parameters for all features—such as thin walls, thick sections, overlaps, and others—remains a critical challenge, in addition to balancing material performance and process productivity. To address this, generalized parameter sets are created to meet as many priorities and geometries as possible. There are still significant opportunities for further refinement of specific parts, geometries, and applications.

Moreover, even for the same material, optimized parameters are continually shifting as manufacturers develop new machine models and technologies, and as processes progress. Deb Roy et al. [29] have provided an excellent review detailing the mechanisms behind various build-related issues, including porosity, residual stress, and preferential evaporation, as well as the grain morphology and texture formed during the LPBF process.

LPBF has several advantages over other AM technologies. One of the primary benefits is the ability to create parts with complex geometries that would be difficult or impossible to produce using traditional manufacturing techniques. LPBF also allows for the creation of parts with variable density, which is useful in applications where weight reduction is critical. Another advantage is the high accuracy of the process, which allows for tight tolerances and intricate details to be achieved. Additionally, the LPBF process can produce parts from a wide range of materials, including metals, ceramics, and composites. However, LPBF also has several disadvantages. that must be taken into consideration. The primary disadvantage is the cost of the equipment and materials, which can be significantly higher than traditional manufacturing processes. Another disadvantage is the limited size of the parts that can be produced using LPBF. The size of the build envelope is typically small, limiting the size of the parts that can be produced. LPBF also has limitations in terms of the surface finish that can be achieved. Additionally, post-processing is required to remove any residual powder from the part, which can be time-consuming and costly.

LPBF is widely used in a variety of industries, including aerospace, medical, and automotive. In the aerospace industry, LPBF is used to produce lightweight, high-strength parts for aircraft engines and other critical components. In the medical industry, LPBF is used to produce orthopaedic implants and prosthetics, which require precise fit and high strength. In the automotive industry, LPBF is used to produce parts for racing vehicles, such as engine components and suspension systems.

LPBF is a promising AM technology with several advantages over traditional manufacturing processes. It allows for the production of complex geometries with high accuracy and strength. While it has some disadvantages, such as high cost and limited build envelope, the benefits make it a valuable tool in various industries. Ongoing research and development will continue to improve the LPBF process, making it an increasingly important manufacturing technology in the future

1.2 Selective Laser Melting (SLM) Process

Selective laser melting (SLM) technology falls under the category of laser powder bed fusion (LPBF) processes, where a layer of metal powder is melted selectively using a laser beam [7,30]. This method presents significant advantages over traditional manufacturing techniques by streamlining production steps, enabling the creation of near-net shapes, and providing enhanced flexibility and efficient material utilization [31]. Complex near-net-shape geometries are generated using CAD data. The surface quality of parts produced via SLM is significantly affected by the process parameters employed [32]. This technology facilitates the production of lattice-structured components, making it particularly appealing for low-volume manufacturing and spare parts. The mechanical properties of these components can be optimized by adjusting the meso-structure through appropriate processing strategies [33]. Recent technological advancements have simplified the process of creating lightweight parts that are both strong and durable. As researchers [30,34] explored the application of AM for achieving high strength and flexibility necessary for reliable medical implants[35,36].

Various metallic materials have been successfully manufactured using SLM, such as maraging steel [37], SS316L [38], titanium alloys [39], INCONEL alloys [7], aluminium alloys [40], and copper alloys [41]. Recognizing the significance of copper alloys in engineering, Ma et al. [42] achieved nearly perfect densification of CuCrZr samples through SLM after refining the process parameters. This led to a remarkable increase in elongation of 67.7% compared to wrought copper alloys with comparable strength characteristics. Popovich et al. [43] discovered a slightly greater ultimate strength and percentage elongation in the direction parallel to the build orientation. Wallis et al. [44] and Guan et al. [45] investigated the micro-hardness and tensile properties of CuCrZr alloys. The relationship between the hardness, porosity, and fractional density of additive manufactured components (which are influenced by Laser Power (LP), Scanning Speed (SS), hatch distance (HD), and powder bed layer thickness (PBLT)) was inaccurately associated with VED [27,46–55].

1.3 Literature survey

The desired mechanical properties and microstructure can be achieved by optimizing processing parameters, which is among the most challenging aspects of AM. Several studies have explored the optimal processing parameters influence on defects, microstructure, and mechanical properties. Here is a summary of the latest literature.

Inconel 718 (IN718) is a prominent nickel-based superalloy used in high-temperature aerospace and automotive applications due to its excellent mechanical properties, such as high tensile strength (UTS), yield strength (YS), and elongation, even in extreme environments. The growing interest in utilizing advanced manufacturing techniques, such as Laser Powder Bed Fusion (LPBF), has led to numerous studies examining the effects of process parameters on the performance of IN718 (see in Table 1.2). This literature survey provides an overview of existing research on how various parameters—including laser power (LP), scanning speed (SS), hatch distance (HD), and powder bed layer thickness (PBLT)—impact the microstructure and mechanical properties of LPBF-processed IN718. The studies highlight the influence of factors such as heat treatment and build orientation, which are critical in optimizing the mechanical performance of IN718 for aerospace-grade applications.

The review of various studies on IN718 and Cu–Cr–Zr processed via different laser-based additive manufacturing techniques, such as SLM, DED, LAM and Direct Metal Laser Sintering (DMLS), highlights significant findings on process parameters and resulting material properties.

For IN718, studies reveal that optimizing laser power (LP), scanning speed (SS), Hatch distance (HD) and powder bed layer thickness (PBLT) can lead to high relative densities (>99%), tensile strengths (up to 1570 MPa), and reduced defects such as lack-of-fusion pores. Heat treatment approaches are essential to mitigate

Laves phases and residual stresses. Fractography identifies LOFs as primary crack initiators, affecting mechanical properties.

For Cu–Cr–Zr, parameter optimization is crucial for achieving high densities and mechanical properties. The use of higher LP (≥300 W) and suitable SS has been effective in producing samples with densities exceeding 99.7%. Specific building angles and controlled hatching strategies influence the repeatability and surface quality of parts.

Overall, tailored parameter sets across both materials have demonstrated the potential for high-performance, dense, and defect-free outputs, emphasizing the need for fine-tuning specific to material and process for achieving desirable properties.

The literature reviewed highlights the critical influence of process parameters such as laser power (LP), scanning speed (SS), and layer thickness (PBLT) on the mechanical properties and microstructure of IN718 and Cu-Cr-Zr components. Key observations show that increasing LP can improve density and tensile strength (UTS), while specific heat treatments can reduce the formation of detrimental phases like Laves. These studies underscore the importance of parameter optimization in achieving high-density, defect-free IN718 parts with improved mechanical properties, providing a basis for further advancements in AM for aerospace-grade materials.

Table 1.2: Literature Survey Table

Material /Process	Parameters Studied	Observations	Reference
IN718 /SLM	LP (W) – 100, 150, 200 SS (mm/s) – 500, 700, 750, 1000, 1500, 2000	YS of 775 MPa, UTS of 1055 MPa, and elongation of 29.5%	Yi et al. [56]
IN718 /SLM	LP of 150 W, SS of 550 mm/s	The tensile strength and elongation were 1570 MPa and 27.3 % respectively	Wang et al. [57]
IN718 /SLM	LP of 180 W, SS of 600 mm/s, PBLT of 30 µm and track width of 0.15 mm	Obtained relative density as 99.10%, the 55-scanning strategy is considered for the optimal manufacturing.	Lu et al. [58]
IN718 / (DED)	LP (W) of 400 – 900, Spot diameter 1 mm, SS in mm/min of 400 – 900	The hardness in range of 260 HV1 ± 10 HV1.	Petrat et al. [59]
IN718 /DMLS	HD 0.11 mm, SS 960 mm/s, LP 285 W, beam offset 0.015 mm and stripe overlap 0.08 mm and PBLT was 40 micrometres	The UTS ranged from 1440 to 1475 MPa and YS ranged from 1253 to 1278 MPa.	Daňa et al. [60]
IN718 /LPBF	LP (W) – 150, 300 SS (mm/s) – 400, 600, 800, 1000, 1200; HD is 60 µm; & PBLT (µm) 30, 50, 70.	Fractography showed the lack-of-fusions(LOFs) were the main crack initiators during the tensile tests, while intergranular cracking due to high-temperature.	Badrossamay et al. [21]
IN718 /LAM	LP of 280 W, PBLT of 40 µm; SS of 960 mm/s; HD of 0.11 mm	Solution heat treatment resulted in reducing these detrimental Laves phases and relief of thermal residual stresses caused by non-homogeneous recrystallization.	Sharma et al. [61]
IN718 /DMLS	LP (W) – 200, 250, 300, 350; SS (mm/s) – 860, 960, 1000; HD is 110 µm and PBLT is 40 µm which kept constant.	The range of process parameters to build nearly defect-free are SS of 860 mm/s to 970 mm/ s; LP of 260W to 300 W.	Kumar et al. [62]
IN718 /SLM	LP of 80 and 480 W. SS of 483 and 2,903 mm/s; HD as 0.1 mm, and PBLT of 30 µm; build preheating temperature of 50 °C, 100 °C, or 150 °C,	A high-density condition was established at 400 W and 967 mm/s. Varying the preheating temperature showed that there was no significant difference in melt pool shape or mechanical properties.	Park et al. [63]
Cu–Cr–Zr /SLM	LP (W) – 270, to 370; SS (mm/s) – 300 to 1000; HD (h) is 120 to 220 µm and PBLT is 40 µm which kept constant.	The optimal parameter set was predicted as 370W of LP, 650 mm/s of SS and 120 mm of HS and the resultant samples exhibit a high density of 8.8761 g/cm3.	Ma et al. [42]
Cu–Cr–Zr /DED	LP (W) – 100, to 400; SS (mm/s) – 400 to 800; HD is 0.11 mm.	That laser power higher than 300 W has to be used in order to manufacture samples with densities exceeding 99.7%.	Jahns et al. [64]
Cu–Cr–Zr /SLM	Factors include the LP, grain size, the PBLT, and the building angle were focused	Concerning the SLM repeatability, better results are obtained in building surfaces facing the powder supplier (i.e., building angle between 60° and 120°).	Caravella et al. [65]
Cu–Cr–Zr /SLM	LP (W) – 160, to 280; SS (mm/s) – 650 to 950; HD is 50 to 80 µm and PBLT is 30 µm.	Optimal process parameters LP is 240 W, SS is 750 mm/s and HD is 0.07 mm. The relative density, UTS and Hv was 98.79%, 347 MPa, 133.9 HV, respectively.	Li et al. [66]

1.3.1 Optimization of Processing Parameters

Jiang et al. [67] employed the concept of equivalent energy density (E_0^*), which depends on several parameters, and focused on three input process variables: LP, SS and HD. By defining these variables, the performance characteristics including relative density (RD), surface roughness (Ra), and hardness (Hv) were assessed. Correlating the performance characteristics with E_0^* is not feasible due to the fact thatE_0^* can yield the same value across different combinations of input process variables, leading to varying performance outcomes.

Sharma et al. [61] observed nearly isotropic characteristics for IN718 by employing LP of 280 W, SS of 960 mm/s, HD of 0.11 mm, and PBLT of 0.04 mm. Sames et al. [68] investigated powder bed temperature and LP affect the mechanical properties and microstructure of IN-718 produced through LPBF. Their results indicated that raising the powder bed temperature (preheat) decreases porosity and improves the fatigue life of the components. However, they also noted that higher temperatures resulted in grain growth and the formation of Laves phase particles, which could weaken the alloy's strength at elevated temperatures.

Deng et al. [69] found that increasing both the LP and SS led to a decrease in the relative density (RD) of the produced component. In contrast, an increase in HD resulted in an increase of the RD. The microstructure of the manufactured parts consisted of predominantly columnar and equiaxed grains, which were more prevalent when high LP and lower SS were employed. Also, the tensile strength and elongation of the component were affected by the processing parameters, with maximum strength observed at intermediate levels of LP and SS, while the highest elongation was recorded at reduced SS.

Onuike et al. [70] explored printing parameters influence the characteristics of IN718 when utilized for repairs through additive manufacturing. They found that optimizing printing parameters, including build orientation and adequate preparation of defect zones, is crucial for achieving high-quality repairs.

Therefore, in any processing technique, it is crucial to thoroughly understand the mechanisms behind defect formation and to identify the process parameters that influence these mechanisms. LPBF process parameters primarily control the thermal input to the material, directly affecting melt pool dimensions, solidification behavior, and thermal history. Consequently, microstructural evolution, defect formation, and mechanical properties are influenced in turn [71]. Any parameter affecting the thermal input conditions and material response is a potential process parameter, leading to the identification of numerous critical parameters. For instance, in the LPBF process, the parameters, mechanisms, and potential defects can be organized as shown in Figure 1.3, modified version with reference to Sun et al. [72].

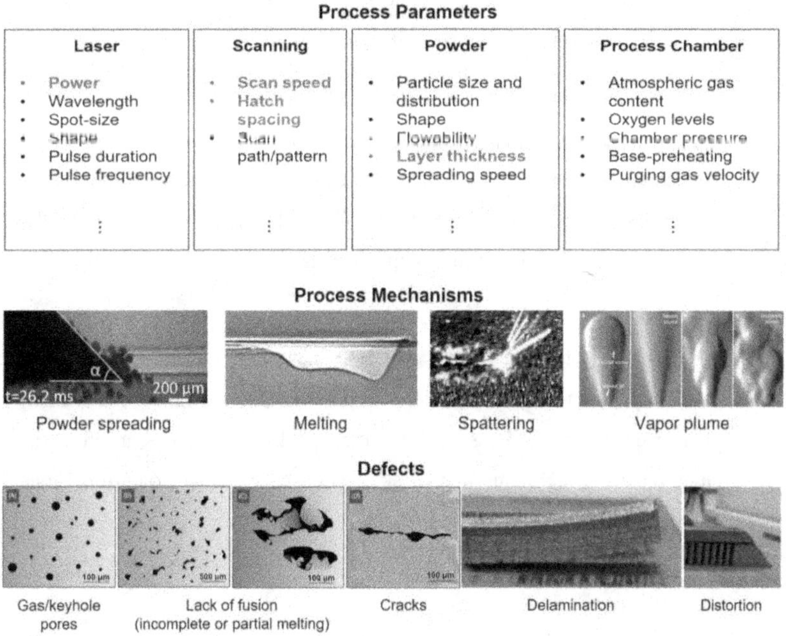

Figure 1.3: LPBF Parameters major influencing the process mechanisms and defect formation

1.3.2 Challenges and Need of Optimization

One of the main challenges of LPBF is the optimization of the process parameters. These parameters, such as laser power (LP), scanning speed (SS), and powder flow rate, significantly affect the quality of the final parts. Therefore, optimizing these parameters is essential for achieving high-quality parts with desired mechanical properties. However, the process is complex, and finding the optimal parameter settings through trial and error basis can be time-consuming and costly. Moreover, the optimal parameters for one material or geometry may not be suitable for another, making optimization a continuous and material-specific task [1,3,29,73].

In recent years, various studies have been conducted to optimize process parameters in LPBF. For example, researchers have used design of experiments (DoE) and response surface methodology (RSM) to optimize the laser power (LP) and scanning speed (SS) for producing parts with minimum porosity and maximum density. In another study, machine learning algorithms were used to predict the optimal parameter settings for LPBF, significantly reducing the time and cost associated with parameter optimization.

Apart from parameter optimization, another challenge of LPBF is the control of part porosity. Porosity can significantly affect the mechanical properties of the final part, reducing its strength and ductility. Therefore, it is essential to minimize porosity during the LPBF process. This can be achieved by controlling the process parameters and using suitable powder bed preparation techniques.

Despite its potential, LPBF still faces several challenges that need to be addressed to enable its wider adoption in various industries. The optimization of process parameters and the control of part porosity are among the critical challenges that need to be addressed. Nevertheless, LPBF remains a promising additive manufacturing technique that offers several benefits over traditional manufacturing methods.

1.3.3 Optimization Techniques

Considering the significance of optimization across various disciplines, a wide array of methods has been utilized on different occasions. Metaheuristic algorithms, for example, have been instrumental in achieving optimal design outcomes. Mirjalili [74], for instance, utilized the anti-lion optimizer (ALO) to solve design problems in structures such as cantilever beams, 3-bar trusses, and gear trains. These metaheuristic algorithms can be further divided into subcategories, including swarm intelligence, evolutionary algorithms, physical algorithms, and hybrid algorithms [75–77]. In structural design, challenges such as welded beam design have been addressed using genetic algorithms (GA), particle swarm optimization (PSO), teaching-learning based optimization (TLBO), and other methodologies [75]. Also, the parameter extraction in photovoltaic models under varying climate conditions, the hybrid arithmetic operations algorithm (HAOA) grounded in the Newton-Raphson method has been implemented [76].

Hairan et al. [77] highlighted the success of combining hybrid metaheuristic algorithms with machine learning models to achieve precise predictions. Manoj et al. [78]utilized the Taguchi-DEAR (Data Envelopment Analysis-based Ranking) approach to investigate how process parameters affect the machining of Al7075 composites that are reinforced with TiB_2 particles in the context of abrasive water jet machining (AWJM). Several different AWJM processes have identified optimal parameters through the use of various algorithms, including cohort intelligence (CI) [79], multi-objective cuckoo search (MOCS) [80], artificial bee colony (ABC) [81], Jay algorithm [82], multi-objective optimization by ratio analysis (MOORA) [83], Gray wolf optimizer (GWO) [84], gravitational search algorithm (GSA) [85], response surface methodology (RSM) [42], artificial neural network (ANN) [86–88], and the Taguchi method alongside evolutionary optimization [89].

Sonawane et al.[90] provided a review encompassing the Taguchi method, GA, Teacher Learning-Based Algorithm (TLBA), Particle Swarm Optimization (PSO), and Grey-based Taguchi method. Taguchi advocated for Signal-to-Noise (S/N) transformation to consolidate scatter in repeated test data into a single value [91]. While S/N ratio transformations address data scatter and yield a single output response value for each test run, the modified Taguchi approach [7,53] engages in multi-response optimization and estimates the anticipated range of the output response for a specified set of process parameters.

Several researchers have adopted the Taguchi method using Minitab software, yet they have not validated the entirety of the generated test data. They typically compare only the optimal solution against the average of the repeated test results. The modified Taguchi approach has been thoroughly assessed regarding its effectiveness [92–98]. A specific study, referenced in [99], made a comparison between modified Taguchi approach [100] and the optimal solutions obtained from a genetic algorithm (GA) and a set of meta-models (EMs) [101], as outlined in [102]. In addition, Chaudhry and Ahmed [102] created a spreadsheet-based GA method specifically designed for optimizing initial aircraft designs. The modified Taguchi approach not only yields optimal solutions but also for developing an empirical relationships for the output response based on input process variables (in non-dimensional form) [103], utilizing the additive law and results from analysis of variance (ANOVA). In this context, the use of Minitab software is not necessary. Also, this method offers a range of estimates for the output response, which is valuable in scenarios involving repeated tests across all levels of input process variables.

To achieve maximum densification and enhanced mechanical properties which builds in the LPBF process, extensive experimentation is essential, focusing on performance indicators and defining a set of SLM processing parameters, including LP, SS and HD.

1.4 Motivation

The adoption of LPBF has significantly expanded the manufacturing capabilities for aerospace-grade materials due to its potential for producing complex, high-performance components with exceptional mechanical properties. However, despite its advantages, optimizing LPBF process parameters remains a challenging task. Variations in process parameters such as LP, SS, HD, and PBLT profoundly influence the resulting densification behavior and mechanical properties. Achieving defect-free and dense components is critical for aerospace applications where precision, reliability, and superior performance are paramount.

The existing literature underscores that empirical optimization methods, including traditional volumetric energy density (VED) calculations, fall short in providing comprehensive insights into the complex interplay between process parameters and resulting material performance. Notably, studies have shown that defects like porosity, lack of fusion (LOF), and inadequate melt pool formation persist, even when energy densities are optimized. Furthermore, the mechanical outcomes are highly dependent on not just the primary parameters but also orientation and post-processing conditions like heat treatment, which are essential to producing properties that surpass those achieved through conventional manufacturing.

To address these challenges, this research aims to explore a systematic optimization approach using a modified Taguchi methodology. The goal is to refine the process parameter landscape for key aerospace materials and defect susceptibilities under LPBF. By establishing empirical relationships between process parameters and performance metrics (e.g., relative density, surface roughness, hardness, and strength), this study seeks to bridge the gap in existing methodologies and contribute to the field by enabling the production of defect-free, high-density components tailored for aerospace applications.

This motivation arises from the demonstrated limitations in current parameter optimization strategies, which often fail to address complex defects and do not fully encapsulate the factors influencing mechanical properties. The research presented herein leverages a comprehensive, data-driven approach through a modified Taguchi method to optimize LPBF process parameters beyond traditional energy density metrics, offering a path toward robust, reproducible results and superior material performance in the aerospace sector.

1.5 Research Gaps

The LPBF technique has attracted significant attention in the production of aerospace-grade materials due to its ability to manufacture complex geometries and high-performance parts. Despite advancements in optimizing LPBF process parameters, challenges remain in achieving maximum densification and enhanced mechanical properties, particularly with materials such as SS316L, IN718, and CuCrZr. The following research gaps have been identified.

- While previous studies have established optimal processing conditions, persistent issues such as irregular porosity, lack of fusion (LOF), and cracking suggest an incomplete understanding of the underlying mechanisms involved.
- Moreover, common metrics like equivalent energy density (E_0^*) and volumetric energy density (VED) fall short in thoroughly explaining the relationship between processing parameters and material performance.
- Additional areas that require further investigation include the influence of build orientation, the interplay between processing parameters and mechanical properties, as well as the effects of post-processing techniques on minimizing defects.

Addressing these research deficiencies is crucial for the progress of LPBF technology, particularly in aerospace applications where material performance is critical. There is a need to bridge these gaps by investigating alternative methods for process optimization, deepening the comprehension of defect formation, and assessing the viability of a comprehensive optimization framework applicable across various materials.

1.5.1 Problem Statement

In LPBF, achieving maximum densification and enhanced mechanical properties for aerospace-grade materials, such as SS316L, IN718, and CuCrZr, remains challenging. Despite optimization efforts, current methodologies—including the widely used volumetric energy density (VED)—fall short in accurately correlating process parameters (LP, SS, HD, and PBLT) with key performance indicators (density, and mechanical strength). Preliminary findings reveal persistent issues like lack of fusion (LOF), porosity, and suboptimal melt pool formation, especially under specific conditions, which compromise material integrity and reliability. This study seeks to refine LPBF optimization techniques by employing a modified Taguchi methodology [104,105] to establish a more resourceful, empirical relationship between process parameters and performance outcomes, thus advancing high densification manufacturing capabilities for aerospace applications.

1.6 Objectives

Objective-I: To identify the ideal SLM process parameters for producing maximum densification and enhanced mechanical properties in samples for aerospace-grade materials (such as SS316L, IN718, and CuCrZr).

Objective-II: To evaluate the relevance of energy density in relating to the performance characteristics of aerospace-grade materials.

Objective-III: To define the optimal SLM process parameters through the modified Taguchi method for attaining superior strength properties of aerospace-grade materials.

1.7 Organization of Thesis

The thesis is structured into five distinct chapters. Chapter 1 provides an overview of the applications and processing involved in LPBF, along with a literature review that identifies research gaps and outlines the study's objectives.

Chapter 2 focuses on the materials and methodology utilized in LPBF.

Chapter 3 delves into theoretical considerations, offering an in-depth discussion on energy density and the Modified Taguchi approach, with its effectiveness demonstrated through a series of case studies.

Chapter 4 presents an experimental investigation aimed at analysing the microstructure and mechanical properties of aerospace-grade materials (such as SS316L, IN718, and CuCrZr).

Finally, Chapter 5 summarizes the findings and discusses potential avenues for future research.

1.8 Summary of Chapter-1

This research explores the optimization of process parameters in Laser Powder Bed Fusion (LPBF) for aerospace-grade materials, focusing on SS316L, IN718, and CuCrZr. The study aims to achieve maximum densification and enhance mechanical properties critical for aerospace applications. Traditional methods, like volumetric energy density (VED), often fall short in correlating process variables—laser power (LP), scanning speed (SS), hatch distance (HD), and powder bed layer thickness (PBLT)—with performance metrics such as density, surface roughness, and strength. Using a modified Taguchi methodology, this thesis establishes empirical relationships between process parameters and performance outcomes. It addresses persistent issues like porosity and lack of fusion, providing insights to optimize LPBF for defect-free, high-density parts and contributing significantly to the field's understanding and applications.

CHAPTER – 2

Materials and Methodology

This chapter focuses on aerospace-grade materials such as SS316L, IN718, and CuCrZr. It outlines the methodology used to examine how SLM process parameters—namely laser power (LP), scanning speed (SS), hatch distance (HD), and powder bed layer thickness (PBLT)—affect densification behavior and mechanical properties. Additionally, it provides an in-depth look at the materials and testing techniques employed to evaluate their performance characteristics.

2.1 Inconel 718 (IN718)

Gas atomized Inconel 718 powder supplied by EOS GmbH (Krailling, Germany) was used to print the samples. The chemical composition of the material is (all in wt%): 53.35 Ni, 19.15 Cr, 4.99 Nb, 3.19 Mo, 1.1 Ti, 0.48 Al and Balance Fe. Measurements were made by inductively coupled plasma atomic emission spectroscopy (ICP-AES). The LPBF machine (EOSINT M280_400W at Central Tool Room & Training Centre (CTTC) in Bhubaneswar) equipped with a single continuous wave fiber laser, was employed to print cubic samples of IN718 with dimensions of 10 mm X 10 mm X 5 mm. All samples were printed on an IN718 substrate pre-heated to 100°C. The most influential processing parameters including LP, SS were varied independently to examine the formation of defects and microstructures. Using Taguchi's L9 orthogonal array, 9 samples were printed with 3 levels of LP (W) (240, 320, 400) and SS (mm/s) (300, 850, 400) at constant HD of 0.11 mm and PBLT of 0.04 mm and an empirical relationship has been established. In order to validate the developed empirical relation, more samples were printed at varying laser power (levels-240, 320, 400, all in W), SS (levels-300, 410, 520, 630, 740, 850, 960, 1070, 1180, 1290, 1400, all in mm/s) at constant HD of 0.11 mm and PBLT of 0.04 mm.

The builds were prepared in a chamber purged with argon gas, and oxygen content was maintained below 0.1 % to control gas pickup. To minimize the thermal stresses during the building process, a meander scanning pattern was adopted for each layer deposition. The scanning direction was rotated by 67° between the adjacent deposition layers. The scan strategy is the tool path that the lasers take during part making.

In order to examine the effect of VED on the density, more test samples were printed at a VED of 67.5 J/mm^3, by varying LP and SS with HD of 0.11 mm and PBLT of 0.04 mm. In addition, a few more samples were printed at a VED of 67.5 J/mm^3 varying LP and HD with SS of 960 mm/s.

After fabrication, the samples were removed from the substrate without performing any stress relieving operation. Density measurements of the printed samples were conducted via Archimedes' method, pursuant to ASTM B962-17. The microstructures were examined using an optical microscope (OM) (OLYMPUS GX71) and scanning electron microscope (SEM) (Carl Zeiss EVO 50) to study the nature of defects and microstructures. The samples were cross sectioned parallel to the build direction (XZ) and perpendicular to the build direction (XY), with the z-axis along the build direction. For metallographic examination, the samples were polished using conventional metallography practices, using a series of emery papers and final polishing was performed using diamond paste of one micron size. The freshly polished specimens were etched using an etchant consisting of hydrochloric acid (HCl), copper chloride ($CuCl_2$), ethanol and nitric acid (HNO_3) for 20-30 seconds.

The freshly polished and etched samples were subjected to hardness measurements. The process for determining Vickers hardness in XY and XZ planes is as follows. In XZ plane a total of 20 and in XY plane a total 16 Hardness values were obtained from four equidistant lines along the X direction, with each line contributing 5 and 4 values, respectively. A load of 300 gf was applied for 13 seconds dwell time and measurements were obtained.

To investigate the impact of AM process parameters on the microstructure and mechanical properties of IN-718, the following experimental configuration was utilized. SLM AM IN-718 cubes with dimensions of 25 x 30 x 40 mm were produced using LPBF with different LP (300 W, 360 W, and 400 W) and SS (520 mm/s, 960 mm/s, and 1400 mm/s), while keeping the HD (0.11 mm) and PBLT (0.04 mm) constant. Each successive printing layer was rotated by 67°. The samples were subjected to heat treatment as per ASTM 3055 [106], as mentioned in Table 2.1.

Each cube was cut into two smaller pieces, as shown in Figure 2.1. These pieces was subjected to the heat treatment (Table 2.1) as per ASTM 3055. Density measurements of the printed samples were conducted via Archimedes' method, pursuant to ASTM B962-17 [7]. The microstructure, identifying defects like porosity, cracks, and lack of fusion, was examined using an optical microscope, specifically the Zeiss Lab A1 AX10 model equipped with a DHS micro cam and Image analyser software. After heat treatment, each piece was further machined to create tensile test specimens with the dimensions specified in Figure 2.1, in two distinct orientations (XZ (90°) direction and the XY (90°) direction). Samples were subjected to tensile testing at both room temperature and 650 °C, to understand the impact of testing temperature on mechanical properties. Tensile testing, with strain rate in accordance with ASTM E8 standards, was conducted using a Zwick Roell All-round Line Z250 SR machine, providing comprehensive data acquisition at both ambient and elevated temperatures.

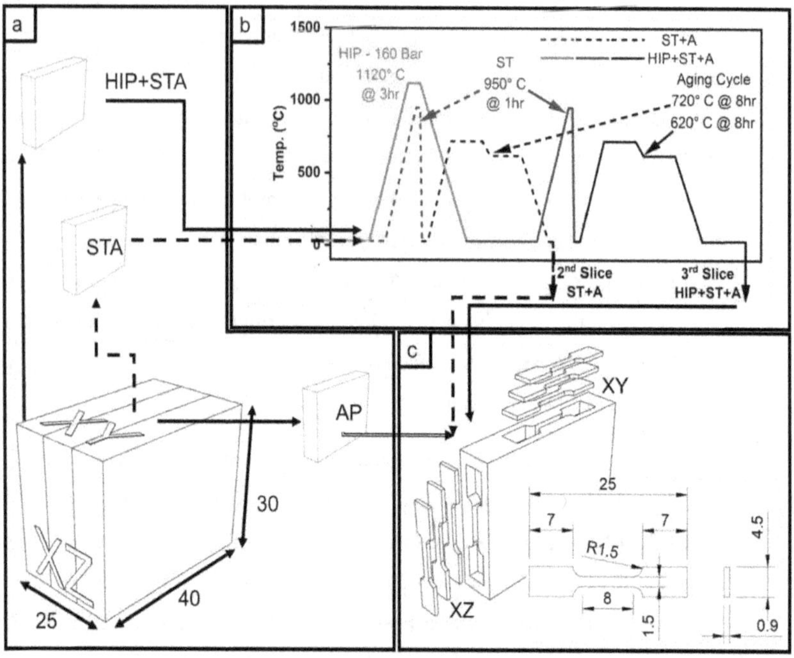

Figure 2.1: Experimental Schematic representation for SLM AM IN718 Specimens [107]

Table 2.1: Heat Treatment Designation for SLM samples [107]

Heat Treatment Nomenclature	Heat Treatment Details
As Printed (AP)	As Printed - No heat treatment
Solution Treatment and Aging (STA)	Solution treatment at 980° C for 1 hr followed by water quenching. Followed by Double aging at 720° C for 8 hr followed by F/C at 55° C/hr to 620° C. Holding at 620° C for 8 hr followed by A/C
Hot Iso-Static Pressing (HIP)	Hot iso-static pressing at 1120° C for 3 hr at 160 bar of pressure followed by Furnace Cooling

The microstructure characterization and mechanical testing of the material were conducted through a series of precise methodologies. Initially, an optical microscope, the Zeiss Lab A1 AX10 model, equipped with DHS micro cam and Image analyser software, was employed to scrutinize the microstructure, identifying defects like porosity, cracks, and lack of fusion. A stereo microscope was utilized for detailed examination of tensile and impact tested specimens. Furthermore, fractography analysis was carried out using the Zeiss Sigma HD Field Emission SEM, offering exceptional resolution and equipped with an energy-dispersive X-ray spectroscopy (EDS) detector. Tensile testing, adhering to ASTM E8 standards, was executed using the Zwick Roell All-round Line Z250 SR Materials Testing Machine, providing comprehensive data acquisition at both ambient and elevated temperatures. These meticulous procedures ensure a thorough understanding of the material's properties and behaviour under various conditions.

To investigate the impact of AM process parameters on the microstructure and mechanical properties of IN-718 the following experimental configuration was utilized. The experimental setting, specimens were additively manufactured using Selective Laser Melting (SLM) technique in four distinct batches by the same printer at different time intervals, allowing us to determine the repeatability of their properties. The effect of printing orientation by printing the samples in four distinct orientations: X (0°), Y (0°), I (45°), and Z (90°) relative to the build plate (see figure 2.2). Samples were subjected to tensile testing at both ambient temperature and 650 °C. This allowed us to understand how orientation, stress relieving, stress relieving followed by solution treatment, and stress relieving followed by solution treatment followed by age hardening, affected the microstructure of IN-718. The blocks were cut into thinner pieces and subjected to a variety of heat treatment conditions. By combining the findings of both experimental designs, thoroughly comprehend the intricate interaction between printing parameters and microstructure and ultimately affect the mechanical properties of IN-718.

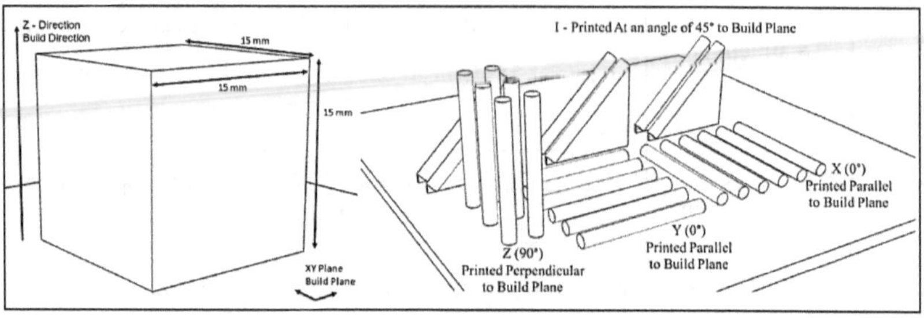

Figure 2.2: Cube Dimension and Print Orientation of samples

In previous study [7] the broad process parameters has been considered for densification, by using those developed empirical relations and optimal parameters made the narrow down of process parameters for the mechanical properties. Using Archimedes principle, the density of the sample can be calculated based on the difference of weight measurements using Equation (2.1) [108].

$$\rho = (\rho_{Fluid} - \rho_{Air}).\frac{mass_{Air}}{mass_{Air} - mass_{Fluid}} + \rho_{Air} \qquad (2.1)$$

Lowest density is observed at both lowest and highest volumetric energy density (VED) while highest density is observed at VED of 85-95 J/mm^3 of VED, which is similar to what was reported in literature (Badrossamay et al.[21]).

2.2 Copper – Chromium – Zirconium alloy (CuCrZr)

The range of SLM process parameters considered were in such the missing range of the parameters fulfil the gap by considering these: 320 W $\leq$ LP $\leq$ 370 W; 610 mm/s $\leq$ SS $\leq$ 720 mm/s; 0.1 mm $\leq$ HD $\leq$ 0.18 mm; and 30 μm $\leq$ PBLT $\leq$ 60 μm. The chemical composition of CuCrZr alloy powder (wt.%): 0.86 Cr, 0.16 Zr, 0.01 Si, 0.0061 S and P, 0.03 Fe and balance Cu. An EOS SLM machine (equipped with a 400 W fibre laser) is used to print 81 cuboidal specimens (10 × 10 × 31 mm) (See Figure 2.3) at M/S Objectify technologies, Delhi. The range of process parameters was selected to compensate for the lack of available literature in the specific regime. Figure 2.4(a) illustrates a detailed view of the 3D printed specimen, indicating the build direction and scanning plane. The XY plane denotes the laser beam's scanning plane, while the XZ plane is orthogonal to it. During SLM process, argon gas is introduced to the build chamber, effectively maintaining low oxygen content (<0.1% volume fraction).

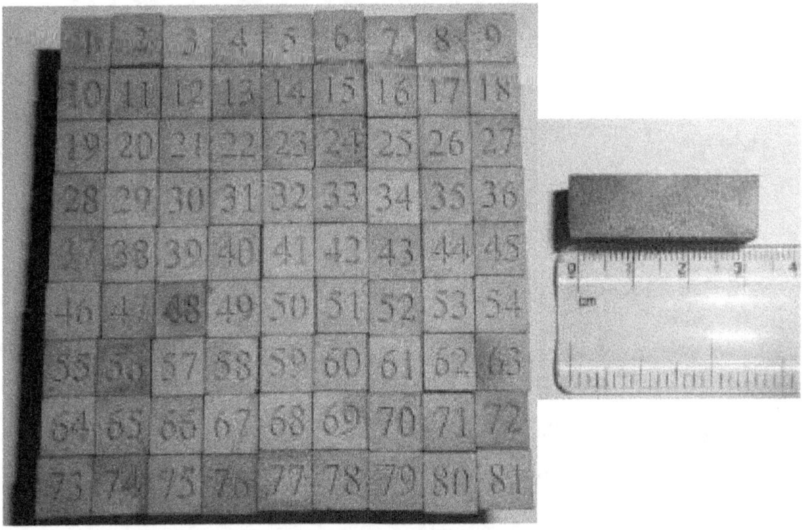

Figure 2.3: 3D printed samples

Figure 2.4(b) illustrates a photomicrograph of CuCrZr spherical powder particles that are encircled by several satellites measuring less than 2 μm, which were analysed employing a Carl Zeiss Sigma HD model Field Emission Scanning Electron Microscope (FESEM). The powder utilized in the current investigation was synthesized through gas atomization, X-ray diffraction (XRD) analysis was conducted for the purpose of phase characterization (see Figure 2.4 (c)) at a scanning velocity of 1°/min, with the resulting spectrum of both the powder and printed specimens in the XY(as printed 1) and XZ (as printed 2) planes being presented in Figure2.4(d).

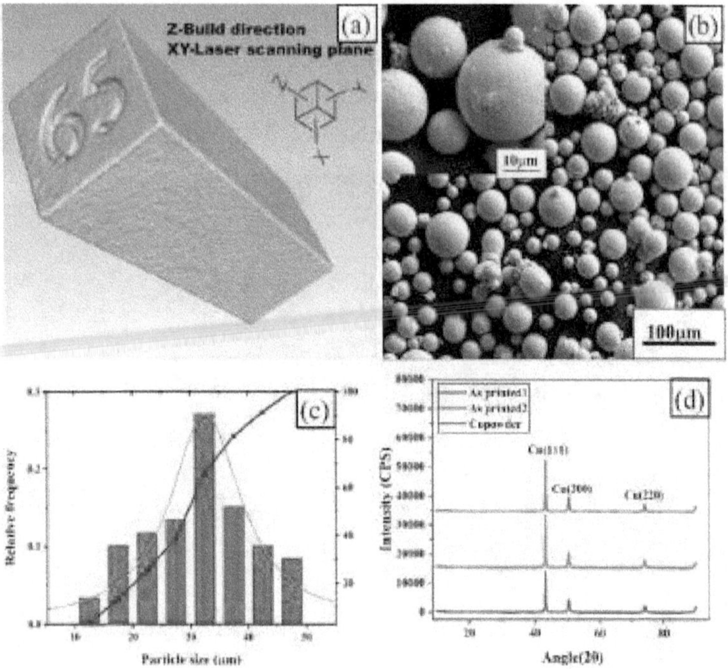

Figure 2.4: (a) Photograph of CuCrZr additively manufactured specimens (b) scanning electron micrograph of the CuCrZr powder with higher magnification image in the inset; (c) plot of the particle size analysis and (d) XRD spectrum of the CuCrZr powder as well as printed specimens in XY (as printed-1) and XZ (as printed-2) planes.

The as-printed specimens were extracted from the substrate without undergoing any thermal treatment. The density of as-printed specimens were measured using Archimedes' principle (ASTM B962-17). For optical microscopic assessment, the specimens underwent a thorough polishing procedure utilizing SiC emery papers with grit sizes of 320, 400, 600, and 800, concluding with a final polishing stage employing one-micron diamond paste. Freshly polished specimens were subjected to chemical etching utilizing a solution comprised of 0.8g $FeCl_3$, 2.5ml HCl, 10ml Glycerine, and 6ml H_2O. Defects and microstructural features were analysed employing an Olympus make (Model GX71). The specimens underwent three distinct thermal processing treatments: direct age hardening (DAH), solution treatment and aging (STA), and a brazing cycle as indicated in Table 2.2. In this context, the brazing cycle refers to the thermal processing sequence that the CuCrZr alloy undergoes during the brazing procedure with a steel outer shell on the divergent segment of the rocket nozzle.

The heat treatment divided into 2 sections, one for direct age hardening, one for Solution treatment (see Figure 2.5) and aging, and finally one cycle for brazing and aging. After heat treatment, the CuCrZr is subjected to a brazing cycle to analyse the printed CuCrZr is responding to mechanical loads. Braze cycle simulation was performed to analyse its mechanical properties (See Figure 2.6).

Table 2.2: Sample designation and corresponding heat treatment conditions

Name	**As Printed**	**DAH**	**STA**	**Brazing cycle**
Heat Treatment Condition	Nil	450^0 for 4 hours and air cooled	Solution treatment at 980°C followed by water quenching followed by aging at 475°C for 8 hours followed by air cooling	Heating the specimen to 980°C for 10 minutes with slow furnace cooling to room temperature followed by aging at 510°C for 2 hours followed by air cooling

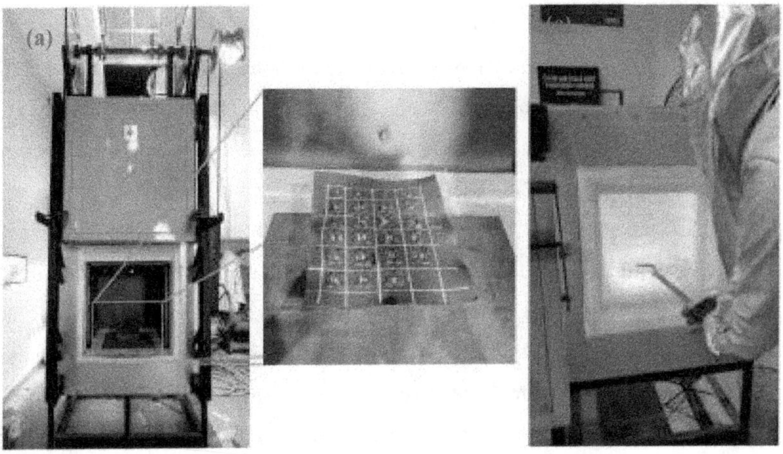

Figure 2.5: CuCrZr samples preparations (a) Ageing furnace with samples loaded, (b) Tensile samples, (c) Solution treatment furnace.

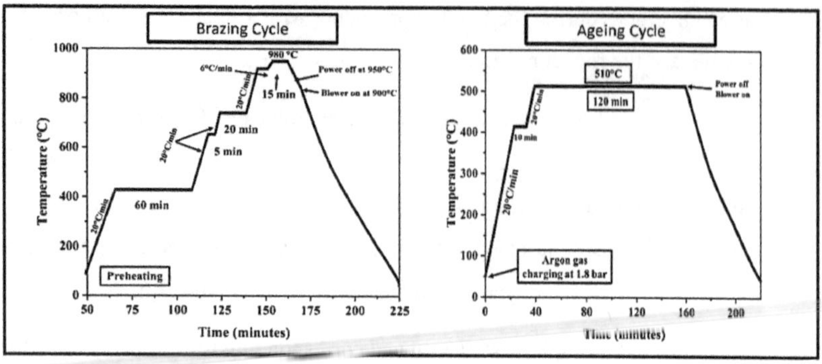

Figure 2.6: Brazing cycle followed by aging cycle for CuCrZr

2.3 Stainless Steel (SS316l)

Comprehensive experiments were conducted by Jiang et al. [67] on SS316L to explore the relationship between process, structure, and properties in the context of the PBF process. They utilized an E Plus metal 3D printer (Model EP 250), equipped with a 400W optical fiber laser that has an 80μm beam diameter, to manufacture specimens measuring 15×15×10 mm via the SLM process. The oxygen content during fabrication was kept below 1000 ppm. The parameters for the SLM process included LP, SS, and HD, with each parameter tested at three different levels. The performance metrics evaluated included surface roughness (Ra), relative density (RD), and hardness (Hv). A surface roughness tester, TR200, was employed to measure the surface roughness of the top surface of each sample. For microstructure characterization of the polished and etched surfaces, OM was applied. Hv measurements were taken using an Everyone MH-6 Vickers indenter, utilizing a 1 kg load and a dwelling time of 20 seconds. The relative density of the additively manufactured samples was assessed with a ZMD-2 electronic densimeter, which calculated the percentage relative to the density (ρ) of SS316L. Data were collected from 27 tests that encompassed a full factorial design of experiments, with three parameters in the SLM process, each varying across three levels.

2.4 Modified Taguchi Method

To produce maximum densification and enhanced mechanical properties for components with aerospace-grade materials (such as SS316L, IN718, and CuCrZr), it is essential to choose the optimum SLM process parameters—specifically, LP, SS, HD, and PBLT. These parameters play a crucial role in enhancing performance characteristics, including Relative Density (RD), yield strength (YS), and ultimate tensile strength (UTS).

For four SLM process parameters, each with three assigned levels, the process designer is required to conduct 81 experiments (i.e., 3^4) by selecting 81 unique sets of process parameters to account for all possible combinations of levels. As the number of levels for the SLM process parameters increases, the number of required experiments also increases, making the task quite challenging, if not unfeasible. Taguchi's design of experiments suggests performing a limited number of experiments while still obtaining comprehensive information. According to Taguchi's orthogonal array (OA), the minimum number of test runs ($N_{Taguchi}$) needed is [109].

$$N_{Taguchi} = 1 + n_p \times (n_l - 1) = 1 + 4 \times (3 - 1) = 9 \qquad (2.2)$$

Here, n_p is the number of process parameters. n_l, is the number of levels assigned to each process parameter. Equation (2.2) suggests that L_9 OA is the more suitable choice. L_9 OA comprises nine distinct sets of process parameters, each of which will undergo testing to evaluate performance characteristics. ANOVA will be performed to derive the 1-mean, 2-mean, and 3-mean values for the performance characteristics associated with each process parameter. The sum of squares (SOS) for the deviations from the mean values relative to the grand mean for each parameter will be calculated. Additionally, the total sum of squares (SOS_T) for the deviations of the measured performance characteristics from their grand mean will be determined. The contribution of each process parameter to the overall performance characteristics is calculated by dividing SOS by SOS_T.

To find the optimum (maximum or minimum) performance characteristics, the levels of each process parameter must be identified based on the maximum or minimum mean values from the ANOVA. Testing will then be conducted for the identified optimal set of process parameters. According to the additive law [109], performance characteristics can be estimated from the mean values in the ANOVA Table by selecting the relevant mean values for the process parameter levels, calculating their deviations from the grand mean, and summing this result with the grand mean. The modified Taguchi method [7,92,96,99,110–112] recommends the application of the additive law to estimate performance characteristics under a full factorial design of experiments. Notably, the 1-mean, 2-mean, and 3-mean values associated with each process parameter correspond to the first, second, and third levels of that parameter. By transforming the levels of each process parameter to range between -1 and +1, the mean values of the performance characteristics can be expressed in a quadratic form, facilitating easier determination of the mean value for a specified process parameter. An empirical relationship for the performance characteristics can be established by substituting these functions along with the grand mean value into the additive law.

As discussed in references [111,112], Chauvenet's criterion is utilized to analyse the output responses from repeated tests, thereby determining the statistically acceptable limits to verify the variability in the repeated test data. The derived allowable limits for the output responses are as follows. For the 3 output responses (η_1, η_2, η_3), $\bar{\eta} = \frac{1}{3}(\eta_1 + \eta_2 + \eta_3)$, is the average value and $\eta_\sigma = \sqrt{\frac{1}{2}\{(\eta_1 - \bar{\eta})^2 + (\eta_2 - \bar{\eta})^2 + (\eta_3 - \bar{\eta})^2\}}$, is the standard deviation. The allowable lower limit, $\eta_{LL} = \bar{\eta} - 1.38 \times \eta_\sigma$. The allowable upper limit, $\eta_{UL} = \bar{\eta} + 1.38 \times \eta_\sigma$. Considering the lower bound value of the performance characteristics, the above analysis gives empirical relationships for the lower bound and similarly, for the upper bound case.

The Taguchi method is particularly effective in determining a set of optimal process parameters for a single objective function. For situations involving multiple objectives, one can identify a set of optimal SLM process parameters by utilizing a simple and reliable multi-objective optimization technique [53,94,112,113].

2.5 Summary of Chapter-2

This chapter outlines the methodology to examine the influence of SLM process parameters—namely laser power (LP), scanning speed (SS), hatch distance (HD), and powder bed layer thickness (PBLT)—on the densification behavior and mechanical properties of aerospace-grade materials such as SS316L, IN718, and CuCrZr.

CHAPTER – 3

Theoretical Considerations

In additive manufacturing, energy density (ED) establishes how much energy can be imparted to a given volume of material, in a given period of time. Utilizing this energy, materials are fused, sintered, or layered to form solid objects. Faster production times and stronger finished parts are often associated with higher ED. Variables must be chosen for the intended AM process, in such a way that the material is completely melted and fused resulting in defect-free part geometries. Overall, the ED in additive manufacturing is a tool that aids in ensuring the properties of the printed object.

3.1 Energy Density

Jiang et al. [67] have considered SS316L and followed the concept of equivalent energy density which is a function of several parameters. Among them, three important input process variables, viz., LP, SS, HD and PBLT were selected and represented as Equation (1). Specifying these three input process variables, the performance indicators were evaluated and plotted with E_0^*. Plots showed large scatter and it was not possible to correlate the performance indicator explicitly with E_0^*. This is due to the fact that E_0^* can have a value for multiple combinations of input process variables with different values of the performance indicator. Several researchers have extensively employed the concept of ED [46–49], while some have specifically referred to it as VED (Equation (3.1)) [50–52,55,114].

$$\mathrm{ED} = \mathrm{VED} = \frac{\mathrm{LP}}{\mathrm{SS}\times\mathrm{HD}\times\mathrm{PBLT}} \tag{3.1}$$

3.1.1 Equivalent Energy Density (E_0^*)

Jiang et al. [67] have considered the equivalent energy density (E_0^*) in terms of process parameters for SS316L in the form [115]

$$E_0^* = \left[\frac{Aq}{2vlh\times\left(\rho C_p\left(T_m - T_0\right)\right)}\right] \tag{3.2}$$

Here, the temperature dependent specific heat capacity, C_p (J/Kg K) is

$$Cp=330.9+0.563T_m-4.015\times10^{-4}\times T_m^2+9.465\times10^{-8}T_m^3 \quad (3.3)$$

The temperature dependent thermal conductivity, λ (W/mK) is

$$\lambda = 11.82+0.0106T_m \quad (3.4)$$

layer thickness, $l = 2.5834\times10^{-5}$ m; beam radius, $R_b = 4\times10^{-5}$ m; thermal diffusivity, $\alpha = 5.38\times10^{-6}$ m²/s; surface absorptivity, A = 0.35; density, ρ = 7980 Kg/m³; melting temperature, T_m=1673 K; initial temperature, T_0 = 333 K.

The equivalent energy density (E_0^*) defined in Equation (3.1) is expressed in terms of the normalized laser power (q^*), scanning speed (v^*), layer thickness (l^*) and hatch distance (h^*) as

$$E_0^* = \frac{q^*}{v^* l^* h^*} \quad (3.5)$$

Here,

$$q^* = \frac{A\times LP}{(R_b\lambda(T_m-T_0))} = 0.220948 \times LP \quad (3.6)$$

$$v^* = \frac{SS\times R_b}{\alpha} = 7.4349\times10^{-3} \times SS \quad (3.7)$$

$$h^* = \frac{HD}{R_b} = 2.5\times10^{-2} \times HD \quad (3.8)$$

$$l^* = \frac{2\times PBLT}{r_b} = 1.292 \quad (3.9)$$

Using equations (3.6) to (3.9) in equation (3.5), the simplified expression for direct evaluation of E_0^* specifying the SLM process parameters (LP, SS and HD) arrived is

$$E_0^* = 920 \times \frac{LP}{SS\times HD} \quad (3.10)$$

3.2 IN718

Factors that can affect the VED include P, SS, HD and PBLT. Generally, IN718 requires efficient melting and solidification during the manufacturing process, which results in strong and dense parts with good mechanical properties. Following the Taguchi's L_9 orthogonal array (OA), the samples were printed at 0.11 mm HD and 0.04 mm PBLT with LP varying from 240-400W and SS varying from 300-1400 mm/s. From the measured density of LPBF processed IN718 samples, the relative density (Equation 3.11) was calculated using the following relation (in reference to Arvieu et al. [116]).

$$\text{Relative Density} = \left(\frac{\text{Density calculated (Archimedes Principle)}}{\text{Theoritical Density of material}}\right) \times 100 \tag{3.11}$$

The theoretical density of IN718 is taken as 8.2 g/cc. As in [92,94,95,113] Table-3.1 provides the levels of LPBF process parameters (viz., LP, SS) and the fictitious parameters (f_1 and f_2). Table-3.1 gives the test data of the performance indicator (viz., density) of the 9 samples as per the Taguchi's L_9 OA.Analysis of variance (ANOVA) results and the significance of LPBF process parameters are presented in Table-3.2. It is seen that percentage of contribution of LP, SS, and the fictitious parameters $f_1$and $f_2$on the density are 2.56%, 94.07%, 2.31%, and 1.06% respectively. The percentage contribution of $f_1$and f_2 on the density (2.31% and 1.06%) is nothing but the error (%).

The 9 conditions along with their measured density is shown in Table-3.3. Lowest density is observed at both lowest and highest volumetric energy density (VED) while highest density is observed at VED of 85-95 J/mm^3 of VED, which is similar to what was reported in literature [21,107]. Empirical relation is developed for the density (ρ) in terms of LP and SS using the additive law from the mean values of ANOVA [7].

$$\rho = 8.1733 + 0.0223\xi_1 - 0.052\xi_1^2 + 0.1787\xi_2 - 0.242\xi_2^2 \tag{3.12}$$

Here, $\xi_1 = \frac{1}{80}(LP - 320)$ and $\xi_2 = \frac{1}{550}(SS - 850)$.

Table 3.1: Test data of the density of 9 samples as per the Taguchi's L9 OA.

Test run	Levels of parameters				Process parameters		Fictitious parameters		Density, ρ (g/cc)
	LP (W)	SS (mm/s)	f_1	f_2	LP (W)	SS (mm/s)	f_1	f_2	
1	1	1	1	1	240	300	f_{11}	f_{21}	7.618
2	1	2	2	2	240	850	f_{12}	f_{22}	8.145
3	1	3	3	3	240	1400	f_{13}	f_{23}	8.050
4	2	1	2	3	320	300	f_{12}	f_{23}	7.764
5	2	2	3	1	320	850	f_{13}	f_{21}	8.176
6	2	3	1	2	320	1400	f_{11}	f_{22}	8.096
7	3	1	3	2	400	300	f_{13}	f_{22}	7.772
8	3	2	1	3	400	850	f_{11}	f_{23}	8.095
9	3	3	2	1	400	1400	f_{12}	f_{21}	8.080

Table 3.2: ANOVA and the significance of LPBF process parameters on the density

Parameters	1-Mean	2- Mean	3-Mean	SOS	%Contribution
Density,ρ (g/cc); grand mean = 7.9773					
LP (W)	7.9377	8.0120	7.9823	8.40E-03	2.56
SS (mm/s)	7.7180	8.1387	8.0753	3.09E-01	94.07
f_1	7.9363	7.9963	7.9993	7.58E-03	2.31
f_2	7.9580	8.0043	7.9697	3.48E-03	1.06

Table 3.3: Relative density estimates

LP (W)	SS (mm/s)	Test Density g/cc	Eq. 3.12	Estimated Range	
				Lower	Upper
300	520	8.01	7.970	7.910	8.019
300	960	8.10	8.191	8.130	8.240
300	1400	7.99	8.101	8.041	8.150
360	520	7.95	7.977	7.917	8.026
360	960	8.13	8.198	8.137	8.247
360	1400	8.12	8.108	8.048	8.157
400	520	8.01	7.949	7.889	7.998
400	960	8.10	8.170	8.090	8.219
400	1400	8.11	8.080	8.020	8.129

The Modified Taguchi Approach was utilized for four process parameters, each set at three different levels, resulting in a minimum of nine required test runs ($N_{Taguchi}$ = 9) [7, 53]. This method was applied to predict the range of output responses, including elongation, 0.2% proof stress or YS, and UTS. The statistical analysis encompassed both samples that were as-printed (XY & XZ) and samples that were AP followed by solution-treated age hardening (XY & XZ). Tables 3.4 and 3.5 display the statistically permissible limits (lower bound (LB) and upper bound (UB)) for the output responses, utilizing Chauvenet's criterion [111]. This analysis includes the average (AVE) and standard deviation (SD) of the repeated test data to confirm the acceptability of variations in the results.

Additive law provides the estimates of the performance indicators from the mean values of ANOVA [7]. The evaluation procedure for the estimates of the performance indicator is explained below. Let ρ be the density (performance indicator) and $\hat{\rho}$ is its estimate for the LPBF process parameters (LP_i, SS_j) varying the subscripts i, j from 1 to 3. $\rho(LP_i)$ and $\rho(SS_j)$ are designated as the mean values of ρ to the ith level of P and the jth level of SS. The grand mean of ρ is designated by ρ_{mean} for the 9 test runs. The test data is found to be within the estimated range. Empirical relation is developed for the Elongation (%), YS (MPa) and UTS (MPa) in terms of LP and SS for both XY and XZ build direction using the additive law from the mean values of ANOVA.

Table 3.4: XY range estimates by using Chauvenet's criterion

As Printed									
Test run	**Parameters**		**% El**		**YS (MPa)**		**UTS (MPa)**		
	LP (W)	**SS (mm/s)**	**LB**	**UB**	**LB**	**UB**	**LB**	**UB**	
1	300	520	32	33	619	831	1043	1132	
2	300	960	32	34	759	828	1062	1171	
3	300	1400	21	35	662	835	1058	1159	
4	360	520	30	34	580	782	1012	1153	
5	360	960	34	38	776	809	1085	1194	
6	360	1400	23	38	500	907	1051	1197	
7	400	520	23	34	661	901	996	1196	
8	400	960	29	41	722	811	1030	1266	
9	400	1400	34	35	672	846	1025	1206	
As Printed and Solution treated aged									
Test run	Parameters		% El		YS (MPa)		UTS (MPa)		
	LP (W)	SS (mm/s)	LB	UB	LB	UB	LB	UB	
1	300	520	15	27	929	1305	1421	1500	
2	300	960	19	20	1308	1334	1461	1592	
3	300	1400	11	20	1110	1430	1489	1536	
4	360	520	14	22	1098	1438	1395	1484	
5	360	960	22	22	1307	1336	1455	1633	
6	360	1400	14	28	910	1450	1445	1563	
7	400	520	14	14	1123	1314	1329	1438	
8	400	960	17	21	1140	1333	1445	1522	
9	400	1400	15	22	1171	1348	1518	1581	
As Printed and Hot iso-static pressing									
Test run	Parameters		% El		YS (MPa)		UTS (MPa)		
	LP (W)	SS (mm/s)	LB	UB	LB	UB	LB	UB	
1	300	520	16	31	940	1385	1442	1496	
2	300	960	20	21	1154	1225	1445	1447	
3	300	1400	21	23	891	1357	1360	1478	
4	360	520	21	25	1068	1448	1403	1501	
5	360	960	20	21	1139	1225	1446	1447	
6	360	1400	17	30	691	1381	1308	1494	
7	400	520	21	27	1130	1235	1465	1485	
8	400	960	18	30	1094	1264	1428	1489	
9	400	1400	17	29	1166	1340	1379	1490	

Table 3.5: XZ range estimates by using Chauvenet's criterion

As Printed									
Test run	Parameters		% El		YS (MPa)		UTS (MPa)		
	LP (W)	SS (mm/s)	LB	UB	LB	UB	LB	UB
1	300	520	36	39	470	724	927	1043
2	300	960	33	37	552	780	994	1104
3	300	1400	27	35	683	756	946	1105
4	360	520	33	43	621	660	978	1077
5	360	960	35	43	579	805	1006	1126
6	360	1400	27	38	550	800	976	1195
7	400	520	34	39	654	713	967	1120
8	400	960	33	45	538	897	949	1105
9	400	1400	26	36	555	775	965	1161
As Printed and Solution treated aged								
Test run	Parameters		% El		YS (MPa)		UTS (MPa)	
	LP (W)	SS (mm/s)	LB	UB	LB	UB	LB	UB
1	300	520	12	28	1095	1156	1294	1427
2	300	960	18	20	1181	1315	1431	1512
3	300	1400	15	23	902	1371	1359	1477
4	360	520	17	22	1105	1243	1370	1401
5	360	960	23	23	1282	1282	1515	1515
6	360	1400	16	26	1101	1409	1357	1509
7	400	520	18	22	1166	1183	1342	1420
8	400	960	13	26	1175	1416	1398	1452
9	400	1400	12	21	1161	1471	1393	1511
As Printed and Hot iso-static pressing								
Test run	Parameters		% El		YS (MPa)		UTS (MPa)	
	LP (W)	SS (mm/s)	LB	UB	LB	UB	LB	UB
1	300	520	25	26	1121	1401	1388	1482
2	300	960	22	26	1128	1244	1427	1473
3	300	1400	18	23	1078	1282	1390	1448
4	360	520	21	30	1105	1376	1374	1504
5	360	960	25	31	1094	1271	1402	1493
6	360	1400	19	20	1042	1227	1390	1450
7	400	520	21	26	1242	1258	1431	1535
8	400	960	25	28	1167	1214	1367	1485
9	400	1400	18	25	1172	1228	1371	1484

3.3 CuCrZr

Table-3.6 gives the levels of print parameters (LP, SS and HD), whose influence on LPBF fabricated samples will be highlighted below. Table-3.6 introduces an additional or fictitious parameter (f) and which gives the measured density for the 9 sets of SLM parameters according to L9 OA.

Table 3.6: Measured density of samples with layer thickness for the 9 test sequence numbers according to Taguchi's L_9 OA and accepted range of density (g/cc) as per the Chauvenet's criterion

Test SNo.	SLM process parameters				Measured density (g/cc)			Estimated range of density (g/cc)	
	LP (W)	SS (mm/s)	HD (mm)	f	PBLT (μm)				
					30	45	60	From	To
1	320	610	0.1	f1	8.421	8.741	8.747	8.371	8.802
2	320	655	0.14	f2	8.687	8.596	8.314	8.141	8.634
3	320	720	0.18	f3	8.487	8.499	8.345	7.873	8.481
4	345	610	0.14	f3	8.726	8.280	8.261	8.320	8.697
5	345	655	0.18	f1	8.647	8.538	8.420	8.087	8.553
6	345	720	0.1	f2	8.163	8.606	8.693	8.485	8.811
7	370	610	0.18	f2	8.744	8.580	8.602	8.228	8.613
8	370	655	0.1	f3	8.063	8.356	8.110	8.570	8.804
9	370	720	0.14	f2	8.702	8.403	8.182	8.365	8.709

As in [111], Chauvenet's criterion is applied to the output responses of the repeated tests in Table-3.6 and determined the statistically allowable limits to verify the acceptance of scatter in the repeated test data. The allowable limits for the output responses arrived as follows. Table-3.6 gives the statistically allowable limits for the repeated test data. However, the output responses are within the statistically allowable limits.

Chauvenet's criterion is used to detect statistically unacceptable values (outliers) from repeated tests [92]. In the present study, this criterion is applied to test the acceptability of density variation with PBLT. The results in Table-3.6 indicate the acceptability of density variation for different PBLT. Considering the lower and upper bounds of the density and applying the modified Taguchi approach [117,118] to the data in Table-3.6, ANOVA results are presented in Table-3.7 to understand the significance of SLM process parameters on the measured total mean. The percentage contribution of LP, SS, HD and f on the lower bound density (ρ_{LL}) is: 28.35 %, 1.78 %, 69.48 % and 0.39 %, while that of the upper bound density (ρ_{UL}) is: 6.82 %, 2.70 %, 89.21 % and 1.27 % respectively. The percentage contribution of f or the error percentage on ρ_{LL} and ρ_{UL} is 0.39 % and 1.27 % respectively.

Table 3.7: ANOVA results with influence of SLM process parameters

Parameters	Mean Value of Density (g/cc)			SOS	%Contribution
	1st	2nd	3rd		
Lower limit, ρ_{LL}; final mean = 8.2712					
LP (W)	8.128	8.297	8.388	1.04E-01	28.35
SS (mm/s)	8.307	8.266	8.241	6.56E-03	1.78
HD (mm)	8.476	8.275	8.063	2.56E-01	69.48
f	8.275	8.285	8.254	1.45E-03	0.39
Upper limit, ρ_{UL}; final mean = 8.6782					
LP (W)	8.639	8.687	8.709	7.56E-03	6.82
SS (mm/s)	8.704	8.663	8.667	2.99E-03	2.70
HD (mm)	8.806	8.680	8.549	9.88E-02	89.21
f	8.688	8.686	8.661	1.40E-03	1.27

The estimates of the density (ρ) from the mean values of ANOVA in Table-3.6 can be obtained using the additive law [119].

$$\hat{\rho} = \rho(LP_i, SS_j, HD_k) = \rho(LP_i) + \rho(SS_j) + \rho(HD_k) - 2 \times \rho_{mean} \quad (3.13)$$

Here $\hat{\rho}$ be the estimate for the specified SLM process parameters (LP_i, SS_j, HD_k), with subscripts i, j, and k are the levels, which are ranging from 1 to 3. $\rho(LP_i)$, $\rho(SS_j)$ and $\rho(HD_k)$ are the mean values of ρ corresponding to the ith level of LP, j^{th} level of SS, and k^{th} level of HD. $\rho_{mean}$is the grand mean of ρ for the 9 test runs.

3.4 SS316l

Jiang et al. [67] have performed extensive experiments to understand the relation of process-structure-property for PBF process. E Plus metal 3D printer (Model EP 250, having 400W optical fiber laser with 80μm beam diameter) was used to print specimens with dimensions 15 mm × 15 mm × 10 mm through SLM process. The oxygen content was maintained below 1000 ppm. The SLM process parameters are LP, SS, and HD. Three levels are assigned to each process parameter. The performance indicators are surface roughness (Ra), relative density (RD), and hardness (H_v). Surface roughness tester-TR200 was used for measuring Ra of each sample on the top printed surface. Optical Microscopy (OM) was utilized for the microstructure characterizations of the polished and etched surfaces. The hardness was measured using Everyone MH-6 Vickers indenter (with 1 kg. load and 20 s dwelling time). ZMD-2 electronic densimeter was used for the relative density of the additively manufactured samples and evaluated its percentage over the RD of SS316L. Data was generated from 27 tests (for 3 SLM process parameters assigning each parameter with 3 levels), which represents the full factorial design of experiments.

Taguchi's design of experiments suggests few experiments. From the acquired test data, there is a possibility to generate the information for the full factorial design of experiments. In the present problem, 3 SLM process parameters are considered assigning 3 levels to each parameter. This corresponds to 27 experiments for the full factorial design of experiments. Taguchi's design of experiments recommends only 9 tests as per L_9 orthogonal array (OA).

To start with, the E_0^* is considered with the L_9 OA data. E_0^* is estimated for the 27 tests in the full factorial design of experiments. For validation of the approach, the estimates will be compared with equation (9). This confirms the adequacy of the Taguchi approach with the exact performance response. Similar procedure will be followed for the other 3 experimentally determined performance indicators. As in [92,94,95,113], without loss of generality, an additional or fictitious parameter (f) is considered in Table-3.8 while presenting the levels of SLM process parameters.

Table 3.8: Levels of SLM process parameters

Process Parameters	Level-1	Level-2	Level-3
Laser Power, LP (W)	206	252	298
Scanning Speed, SS (mm/s)	900	1100	1300
Hatch Distance, HD (μm)	75	90	105
Fictitious Parameter, f	f1	f2	f3

Table-3.9 presents the levels of the SLM process parameters (viz., LP, SS, HD and f) as per the Taguchi's L_9 OA and the test data of the performance indicators (viz., equivalent energy density (E_0^*), surface roughness (R_a), Relative density (RD) of the samples, and hardness (H_v)) for the nine test runs. Table-3.9 presents ANOVA results.

Table 3.9: Test Data of performance indicators

Test Run	Levels of parameters				Test Data [67]			
	LP (W)	**SS (mm/s)**	**HD (μm)**	**f**	E_0^*	R_a **(μm)**	**RD (%)**	H_v
1	1	1	1	f1	2.808	9.80	99.35	182.38
2	1	2	2	f2	1.914	11.55	98.85	186.85
3	1	3	3	f3	1.388	11.95	96.84	134.99
4	2	1	2	f3	2.862	8.43	99.28	211.98
5	2	2	3	f1	2.007	6.79	98.99	195.12
6	2	3	1	f2	2.378	6.76	99.15	198.96
7	3	1	3	f2	2.901	4.25	99.17	199.80
8	3	2	1	f3	3.323	3.92	99.24	204.88
9	3	3	2	f1	2.343	7.11	99.22	197.57

Table 3.10: Analysis of variance (ANOVA) and the significance of process parameters

Parameters	1-Mean	2- Mean	3-Mean	(SOS)	% Contribution
E_0^*; grand mean = 2.436					
LP (W)	2.04	2.42	2.86	1.01	34.9
SS (mm/s)	2.86	2.41	2.04	1.01	35.0
HD (μm)	2.84	2.37	2.10	0.83	28.9
f	2.39	2.40	2.52	0.04	1.2
R_a (μm); grand mean = 7.840					
LP (W)	11.10	7.33	5.09	55.31	83.9
SS (mm/s)	7.49	7.42	8.61	2.65	4.0
HD (μm)	6.83	9.03	7.66	7.42	11.3
f	7.90	7.52	8.10	0.52	0.8
RD (%); grand mean = 98.899					
LP (W)	98.35	99.14	99.21	1.38	27.9
SS (mm/s)	99.27	99.03	98.40	1.19	24.0
HD (μm)	99.25	99.12	98.33	1.46	29.6
f	99.19	99.06	98.45	0.92	18.5
H_v; grand mean = 190.281					
LP (W)	168.07	202.02	200.75	2221.75	54.8
SS (mm/s)	198.05	195.62	177.17	782.07	19.2
HD (μm)	195.41	198.80	176.64	855.04	21.1
f	191.69	195.20	183.95	198.89	4.9

It is seen that percentage contribution of laser power (LP), scanning speed (SS), hatch distance (HD) and the fictitious parameter (f) on the equivalent energy are 34.9 %, 35%, 28.9% and 1.2 % respectively; surface roughness (R_a) are 83.9 %, 4 %, 11.3 % and 0.8 % respectively; Relative Density (RD) are 27.9 %, 24 %, 29.6 % and 18.5 % respectively; and hardness are 54.8 %, 19.2 %, 21.1 % and 4.9 % respectively. The percentage contribution of the fictitious parameter (f) on the equivalent energy, surface roughness, relative density and hardness are 1.2 %, 0.8 %, 18.5 % and 4.9 % respectively, which are nothing but the error (%).

Additive law [109] provides the estimates of performance indicators from the mean values of ANOVA presented in Table-3.10. The evaluation procedure for the estimates of the performance indicators is explained below. Let ψ be the performance response and $\hat{\psi}$ is its estimate for the SLM process parameters (LP_i, SS_j, HD_k, f_l) varying the subscripts i, j, k, l from 1 to 3. Designating $\psi(LP_i)$, $\psi(SS_j)$, $\psi(HD_k)$and

$\psi(f_l)$ as mean values of ψ corresponding to the ith level of LP, j^{th} level of SS, k^{th} level of HD and l^{th} level of f. The grand mean of ψ is designated by ψ_{mean} for the 9 test runs. Following the additive law, estimate $\hat{\psi}$ for the specified (LP_i, SS_j, HD_k, f_l) is:

$$\hat{\psi} = \psi(LP_i, SS_j, HD_k, D_l) = \psi(LP_i) + \psi(SS_j) + \psi(HD_k) + \psi(f_l) - 3 \times \psi_{mean} \tag{3.14}$$

For the specified (LP_i, SS_j, HD_k), equation (3.14) reduces to

$$\hat{\psi} = \psi(LP_i, SS_j, HD_k) = \psi(LP_i) + \psi(SS_j) + \psi(HD_k) - 2 \times \psi_{mean} \tag{3.15}$$

3.5 Summary of Chapter-3

This chapter provides the test data as per the Taguchi's L_9 OA (orthogonal array) for aerospace-grade materials, including SS316L, IN718, and CuCrZr. An analysis of variance (ANOVA) is conducted to investigate the effects of selective laser melting (SLM) process parameters—specifically, laser power (LP), scanning speed (SS), hatch distance (HD), and powder bed layer thickness (PBLT)—on the densification characteristics and mechanical properties of SS316L, IN718, and CuCrZr. The methodology for evaluation is detailed through the ANOVA findings related to performance indicator estimates.

CHAPTER – 4

Results and Discussion

4.1 Densification behaviour of Aerospace grade Materials

Samples of aerospace-grade materials (such as SS316L, IN718, and CuCrZr) were made through different SLM process conditions as per the Taguchi's L_9 OA to study the densification behavior, mechanical properties and microstructural analysis.

IN718

Additive law provides the estimates of the performance indicator (viz., density) from the mean values of ANOVA (Table-3.2). The evaluation procedure for the estimates of the performance indicator is explained below. Let ρ be the density (performance indicator) and $\hat{\rho}$is its estimate for the LPBF process parameters (LP_i, SS_j) varying the subscripts $i, j$from 1 to 3. $\rho(P_i)$ and $\rho(SS_j)$ are designated as the mean values of ρ to the ith level of Pand the jth level of SS. The grand mean of ρ is designated by ρ_{mean} for the 9 test runs. Following the additive law, estimate $\hat{\rho}$ for the specified (LP_i, SS_j) is:

$$\hat{\rho} = \rho(LP_i, SS_j) = \rho(LP_i) + \rho(SS_j) - \rho_{mean} \tag{4.1}$$

Table-4.1 provides estimates of the density, $\hat{\rho}$ (g/cc) from the mean values in ANOVA Table-3.2 using additive law (Equation 4.1). Grand mean, $\rho_{mean} = 7.9773$ g/cc. The discrepancy between estimates and the test data is within 1%.

Table 4.1: Estimates of the density from the mean values in ANOVA

Test run	Process parameters		Mean values of density, ρ from ANOVA Table-3.2			Density (g/cc)		Relative Error, R.E. (%)
	LP (W)	SS (mm/s)	(i^{th}, j^{th})	$\rho(LP_i)$	$\rho(SS_j)$	Using Eq.4.1	Test	
1	240	300	(1, 1)	7.9377	7.7180	7.678	7.618	-0.79
2	240	850	(1, 2)	7.9377	8.1387	8.099	8.145	0.56
3	240	1400	(1, 3)	7.9377	8.0753	8.036	8.050	0.18
4	320	300	(2, 1)	8.0120	7.7180	7.753	7.764	0.15
5	320	850	(2, 2)	8.0120	8.1387	8.195	8.176	-0.24
6	320	1400	(2. 3)	8.0120	8.0753	8.110	8.096	-0.17
7	400	300	(3, 1)	7.9823	7.7180	7.745	7.772	0.35
8	400	850	(3, 2)	7.9823	8.1387	8.144	8.095	-0.60
9	400	1400	(3, 3)	7.9823	8.0753	8.080	8.080	0.00

Deviations in density (ρ) from the mean values of the fictitious parameters f_1 and f_2 over the grand mean (ρ_{mean}) were obtained from:

$$\Delta\rho_{ij} = \rho(f_{ij}) - \rho_{mean} \qquad \forall\, i, j = 1 \text{ to } 3 \qquad (4.2)$$

For example, $\Delta\rho_{11} = \rho(f_{11}) - \rho_{mean} = 7.9363 - 7.9773 = -0.0410$. Similarly, other deviations in density can be found from the ANOVA Table-3.2.

Corrections $\Delta\rho_-$ and $\Delta\rho_+$ to the estimates from the additive law (Equation 4.1) were made for obtaining lower and upper bounds as:

$$\Delta\rho_- = min\{\Delta\rho_{11}, \Delta\rho_{12}, \Delta\rho_{13}\} + min\{\Delta\rho_{21}, \Delta\rho_{22}, \Delta\rho_{23}\}$$

= min{-0.0410, 0.0190, 0.0220} + min{-0.0193, 0.0270, -0.0077}

= -0.0410-0.0193 = -0.0603 (4.3)

$$\Delta\rho_+ = max\{\Delta\rho_{11}, \Delta\rho_{12}, \Delta\rho_{13}\} + max\{\Delta\rho_{21}, \Delta\rho_{22}, \Delta\rho_{23}\}$$

= max {-0.0410, 0.0190, 0.0220} + max{-0.0193, 0.0270, -0.0077}

= 0.022+0.027=0.049 (4.4)

Applying the corrections $\Delta\rho_{-} = -0.0603$ and $\Delta\rho_{+} = 0.049$ to the density obtained from the additive law (Equation 4.1), one can find the expected range of the density. Table-4.2 gives the expected range of density for the 9 test runs. The test data is found to be within/close to the lower bound value $(\hat{\rho} + \Delta\rho_{-})$and the upper bound value $(\hat{\rho} + \Delta\rho_{+})$. In fact, additive law (Equation4.1) estimates the density for all combinations of the LPBF process parameters. Empirical relation for the density in terms of LPBF process parameters is required for estimating the density throughout the range of the parameters.

Table 4.2: Expected range of density for the test runs in Taguchi's L9 OA

Test run	**Process Parameters**		**Test density, ρ (g/cc)**	**Estimated density, (g/cc)**		
	LP (W)	**SS (mm/s)**		**Using Eq.4.1**	**LB**	**UB**
1	240	300	7.618	7.678	7.618	7.727
2	240	850	8.145	8.099	8.039	8.148
3	240	1400	8.050	8.036	7.975	8.085
4	320	300	7.764	7.753	7.692	7.802
5	320	850	8.176	8.195	8.135	8.244
6	320	1400	8.096	8.110	8.050	8.159
7	400	300	7.772	7.745	7.685	7.794
8	400	850	8.095	8.144	8.083	8.193
9	400	1400	8.080	8.080	8.020	8.129

Empirical relation is developed for the density (ρ) in terms of LP and SS using the additive law from the mean values of ANOVA.

$$\rho = 8.1733 + 0.0223\xi_1 - 0.052\xi_1^2 + 0.1787\xi_2 - 0.242\xi_2^2 \quad (4.5)$$

Here, $\xi_1 = \frac{1}{80}(LP - 320)$ and $\xi_2 = \frac{1}{550}(SS - 850)$.

Estimations in Table-4.3(a) to 4.3(c) are from the developed empirical relation (Equation 4.5) by varying LP from 240 to 400 W; SS from 300 to 1400 mm/s at constant HD of 0.11 mm and PBLT of 0.04mm. The test data is found to be within the estimated range.

Table 4.3(a): Estimates of the density, by varying SS (300 mm/s to 630 mm/s)

LP (W)	SS (mm/s)	VED (J/mm^3) Eq.-3.1	ρ (g/cc)	RD Eq.-3.11	Estimates of density,ρ (g/cc)		
					Empirical Relation Eq.-4.5	LB	UB
240	300	181.82	7.618	92.90	7.678	7.618	7.727
280	300	212.12	7.613	92.84	7.728	7.668	7.777
320	300	242.42	7.764	94.68	7.753	7.692	7.802
360	300	272.73	7.797	95.09	7.751	7.690	7.800
400	300	303.03	7.772	94.78	7.723	7.663	7.772
240	410	133.04	7.880	96.10	7.801	7.741	7.850
280	410	155.21	7.798	95.10	7.851	7.791	7.900
320	410	177.38	7.848	95.71	7.875	7.815	7.924
360	410	199.56	7.872	96.00	7.874	7.813	7.923
400	410	221.73	7.827	95.45	7.846	7.785	7.895
240	520	104.90	8.107	98.87	7.905	7.844	7.954
280	520	122.38	8.096	98.73	7.955	7.895	8.004
320	520	139.86	8.004	97.61	7.979	7.919	8.028
360	520	157.34	7.983	97.35	7.977	7.917	8.026
400	520	174.83	8.024	97.85	7.949	7.889	7.998
240	630	86.58	8.041	98.06	7.989	7.929	8.038
280	630	101.01	8.122	99.05	8.039	7.979	8.088
320	630	115.44	8.180	99.76	8.063	8.003	8.112
360	630	129.87	8.118	99.00	8.061	8.001	8.110
400	630	144.30	8.111	98.91	8.033	7.973	8.082

Table 4.3(b): Estimates of the density, by varying SS (740 mm/s to 960 mm/s)

LP (W)	SS (mm/s)	VED (J/mm^3) Eq.-3.1	ρ (g/cc)	RD Eq.-3.11	Estimates of density, ρ (g/cc)		
					Empirical Relation Eq.-4.5	LB	UB
240	740	73.71	7.995	97.50	8.054	7.993	8.103
280	740	86.00	7.993	97.48	8.104	8.043	8.153
320	740	98.28	8.207	100.09	8.128	8.068	8.177
360	740	110.57	8.193	99.91	8.126	8.066	8.175
400	740	122.85	8.138	99.24	8.098	8.038	8.147
240	850	64.17	8.145	99.33	8.099	8.039	8.148
280	850	74.87	8.069	98.40	8.149	8.089	8.198
320	850	85.56	8.176	99.71	8.173	8.113	8.222
360	850	96.26	8.253	100.65	8.171	8.111	8.220
400	850	106.95	8.095	98.72	8.144	8.083	8.193
240	960	56.82	8.035	97.99	8.125	8.065	8.174
280	960	66.29	8.096	98.73	8.175	8.115	8.224
320	960	75.76	8.105	98.84	8.199	8.139	8.248
360	960	85.23	8.141	99.28	8.198	8.137	8.247
400	960	94.70	8.165	99.57	8.170	8.109	8.219

Table 4.3(c): Estimates of the density, by varying SS (1070 mm/s to 1400 mm/s)

LP (W)	SS (mm/s)	VED (J/mm^3) Eq.-3.1	ρ (g/cc)	RD Eq.-3.11	Estimates of density, ρ(g/cc)		
					Empirical Relation Eq.-4.5	LB	UB
240	1070	50.98	8.073	98.45	8.132	8.071	8.181
280	1070	59.47	8.054	98.22	8.182	8.122	8.231
320	1070	67.97	8.139	99.26	8.206	8.146	8.255
360	1070	76.47	8.153	99.43	8.204	8.144	8.253
400	1070	84.96	8.098	98.76	8.176	8.116	8.225
240	1180	46.22	8.109	98.89	8.119	8.059	8.168
280	1180	53.93	8.115	98.96	8.169	8.109	8.218
320	1180	61.63	8.109	98.89	8.193	8.133	8.242
360	1180	69.34	8.108	98.88	8.192	8.131	8.241
400	1180	77.04	8.118	99.00	8.164	8.103	8.213
240	1290	42.28	8.087	98.62	8.087	8.027	8.136
280	1290	49.33	8.046	98.12	8.137	8.077	8.186
320	1290	56.38	8.133	99.18	8.161	8.101	8.210
360	1290	63.42	8.133	99.18	8.160	8.099	8.209
400	1290	70.47	8.117	98.99	8.132	8.071	8.181
240	1400	38.96	8.050	98.17	8.036	7.975	8.085
280	1400	45.45	8.153	99.43	8.086	8.026	8.135
320	1400	51.95	8.096	98.73	8.110	8.050	8.159
360	1400	58.44	8.128	99.12	8.108	8.048	8.157
400	1400	64.94	8.080	98.54	8.080	8.020	8.129

Figure-4.1(a) compares the density using additive law and test data. Figure-4.1 (b) shows test data of density in between the estimated range (upper bound and lower bound), which clearly indicates the adequacy of the developed empirical relation (Equation 4.5). From the ANOVA Table-3.2, the maximum density is expected for the set of LPBF processing parameters (P_2, SS_2) in which the subscripts denote the level of the process parameters.

The optimal LPBF process parameters to achieve maximum density are the laser power, $LP_2 = 320$ W and the scanning speed, $SS_2 = 850$ mm/s which correspond to the test run-5 in Table-3.1 of the Taguchi's L_9 OA, and the achieved density, $\rho_{max} = 8.176$ g/cc. The additive law (Equation 4.1) estimates the maximum density, $\rho_{max} = 8.195$ g/cc, and the expected range is from 8.135 to 8.244 g/cc. The test data is within the expected range. Since, percentage contribution of SS is high, the variation of density with scanning speed is shown in Figure-4.1(c) at a constant laser power (320 W). The variation of density with laser power in Figure-4.1(d) is for 850 mm/s scanning speed. All estimates of the density including test data were made for all possible combinations of LPBF process parameters. Variation of measured density with VED considering complete test data is presented in Figure-4.2. The large scatter in density with VED (see Figure-4.2 (a)) is mainly attributed to different sets of LPBF process parameters which yield a given VED value. This should not be treated as an experimental scatter. The dispersion in density for VED of 67.5 J/mm^3 is clearly visible in Figure-4.2 (b).

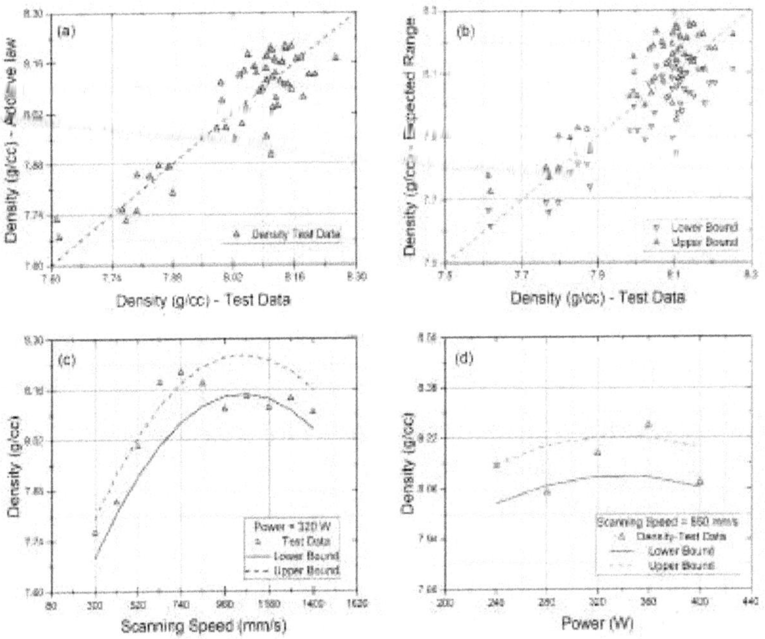

Figure 4.1: (a) comparison of density estimates with the test data; (b) comparison of density test data with estimates; (c) variation of density with SS of 320W LP at HD of 0.11mm and PBLT of 0.04mm and (d) variation of density with LP of 850 mm/s SS at HD of 0.11mm and PBLT of 0.04mm.

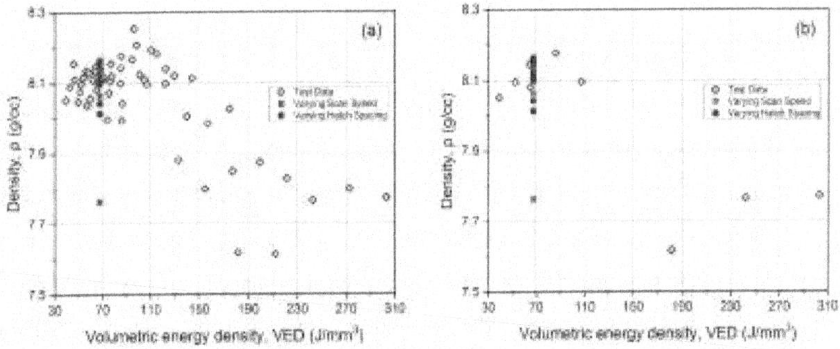

Figure 4.2: (a) variation of density with VED all test data and (b) variation of Density with VED from the data in L_9 orthogonal array.

CuCrZr

Correction to the lower bound estimates of ρ can be obtained from the lower mean value of ρ for fictitious parameter (f) and the total mean value. Similarly, correction to the upper bound estimates of ρ can be obtained from the higher mean value of ρ for fictitious parameter (f) and the total mean value. Applying corrections to the lower and upper bound values of ρ, the range of estimates is arrived. Following [94], empirical relationships for lower density (ρ_{LL}) and upper density (ρ_{UL}) have been developed by using ANOVA results in the additive law, expressed in terms of LP, SS and HD (Using Equation 3.15).

$$\rho_{LL} = 8.2766 + 0.12986\xi_1 - 0.03920\xi_1^2 - 0.03278\xi_2 + 0.00744\xi_2^2 - 0.20634\xi_3 - 0.0062\xi_3^2 \quad (4.6)$$

$$\rho_{UL} = 8.6838 + 0.0347\xi_1 - 0.01291\xi_1^2 - 0.01833\xi_2 + 0.02212\xi_2^2 - 0.12833\xi_3 - 0.00291\xi_3^2 \quad (4.7)$$

Here, $\xi_1 = \frac{1}{25}(LP - 345)$; $\xi_2 = \frac{1}{64350}((SS - 655) \times (3870 - (4 \times SS)))$ and $\xi_3 = \frac{1}{0.04}(HD - 0.14)$

For the full factorial design of experiments, 27 sets of process parameters are possible for all combinations of the 3 levels of the 3 parameters: $\left(\left(\left((LP_i\, SS_j\, HD_k), k = 1,2,3\right), j = 1,2,3\right), i = 1,2,3\right)$, which are given in Table-4.4.

Density values generated using equation 4.6 and 4.7 for the 27 sets of parameters (see Table-4.4) are shown in Figure-4.3. Measured density values in Table-4.4 are within/close to the estimated range, as shown in Figure-4.3. The test sequence number (Sno.) in the two dimensional plot of Figure-4.3 correspond to the set of parameters in Table-4.4. As per the Chauvenet's criterion, scatter in the measured density by varying PBLT from 30 μm to 60 μm is acceptable with 95 % confidence level (assuring 95 samples data out 100). Hence, the influence of PBLT on average density is insignificant. Therefore, most of the densities of specimens printed with different PBLT in 27 experiments lie between the lower and upper limits of density as shown in Figure-4.3. Table-4.4 presents the set of SLM process parameters for each test run. As expected, the measured densities for all 3 layer thicknesses were within the estimated range.

Table 4.4: Density (g/cc) estimates

Sno.	Parameters			Density (g/cc)				
				PBLT (µm)			Estimated Range	
	LP (W)	SS (mm/s)	HD (mm)	30	45	60	From (Eq.4.6)	To (Eq.4.7)
1	320	610	0.10	8.726	8.616	8.418	8.348	8.802
2	320	610	0.14	8.546	8.163	8.345	8.148	8.677
3	320	610	0.18	8.401	8.422	8.062	7.935	8.545
4	320	655	0.10	8.741	8.42	8.42	8.308	8.762
5	320	655	0.14	8.548	8.181	8.195	8.108	8.636
6	320	655	0.18	8.397	8.616	8.192	7.895	8.505
7	320	720	0.10	8.747	8.553	8.389	8.282	8.765
8	320	720	0.14	8.498	8.377	8.063	8.082	8.640
9	320	720	0.18	8.395	8.182	7.954	7.870	8.509
10	345	610	0.10	8.739	8.681	8.421	8.517	8.850
11	345	610	0.14	8.62	8.549	8.356	8.317	8.724
12	345	610	0.18	8.471	8.314	8.141	8.104	8.593
13	345	655	0.10	8.772	8.674	8.508	8.477	8.809
14	345	655	0.14	8.606	8.534	8.406	8.277	8.684
15	345	655	0.18	8.487	8.323	8.15	8.064	8.553
16	345	720	0.10	8.744	8.684	8.516	8.451	8.813
17	345	720	0.14	8.609	8.499	8.286	8.251	8.688
18	345	720	0.18	8.465	8.261	8.11	8.039	8.556
19	370	610	0.10	8.766	8.716	8.546	8.608	8.871
20	370	610	0.14	8.687	8.647	8.515	8.407	8.746
21	370	610	0.18	8.58	8.36	8.322	8.195	8.615
22	370	655	0.10	8.763	8.702	8.596	8.567	8.831
23	370	655	0.14	8.693	8.602	8.504	8.367	8.706
24	370	655	0.18	8.538	8.403	8.28	8.155	8.574
25	370	720	0.10	8.79	8.704	8.511	8.542	8.835
26	370	720	0.14	8.651	8.557	8.404	8.342	8.709
27	370	720	0.18	8.556	8.32	8.122	8.129	8.578

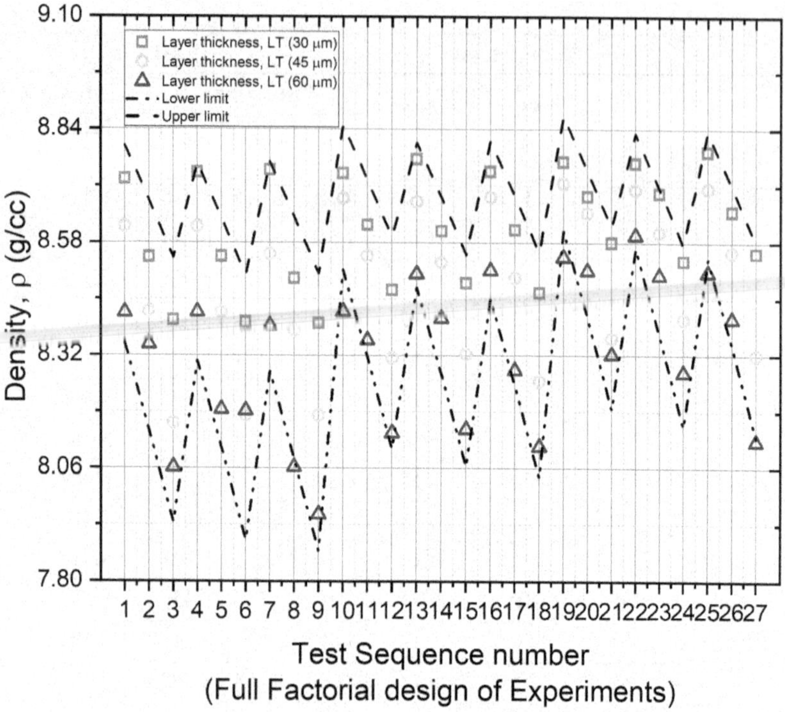

Figure 4.3: Range estimation for the density (g/cc) using PBLT.

SS316L

In case of SS316L, it is seen that percentage contribution of LP, SS, HD and f on the relative density (RD) are 27.9 %, 24 %, 29.6 % and 18.5 % respectively. Figure-4.4 presents the estimates of the relative density including test data (Jiang et al., 2019) [67] for all possible 27 parameters. Empirical relationship (Equation 4.8) is developed for the Relative Density in terms of LP, SS, and HD using the additive law from the mean values of ANOVA presented in Table-4.5 in the form:

$$\rho = 99.4856 + 0.4317\xi_1 - 0.3617\xi_1^2 - 0.4317\xi_2 - 0.1917\xi_2^2 - 0.45673\xi_3 - 0.3267\xi_3^2 \quad (4.8)$$

Here, $\xi_1 = \frac{LP-252}{46}$, $\xi_2 = \frac{SS-1100}{200}$ and $\xi_3 = \frac{HD-90}{15}$.

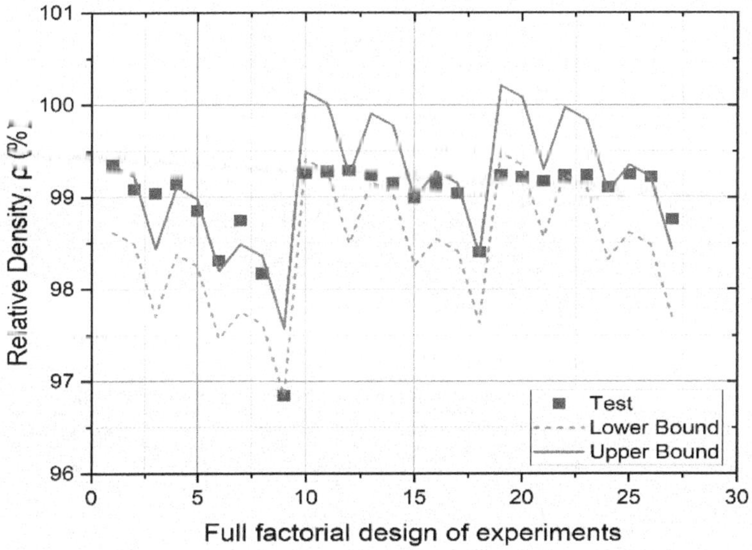

Figure 4.4: Relative density, ρ (%) including test data [67] for Full factorial design of experiments

Table 4.5: Significance of SLM process parameters on relative density of SS316L

Parameters	**1-Mean**	**2- Mean**	**3-Mean**	**SOS**	**% Contribution**
Relative Density (%); grand mean = 98.899					
LP (W)	98.35	99.14	99.21	1.38	27.9
SS (mm/s)	99.27	99.03	98.40	1.19	24.0
HD (μm)	99.25	99.12	98.33	1.46	29.6
f	99.19	99.06	98.45	0.92	18.5

4.2 Mechanical Properties of Aerospace grade Materials

IN718

Estimations are from the developed empirical relation (see Table 4.6) by varying LP from 300 to 400 W: SS from 520 to 1400 mm/s at constant HD of 0.11 mm and PBT of 0.04 mm can be seen in Figure-4.5 and Figure-4.6. Let ψ be the general function for output response, which can be expressed in terms of input variables as in equation 4.9.

$$\psi = \psi_0 + \psi_1\xi_1 + \psi_2\xi_1^2 + \psi_3\xi_2 + \psi_4\xi_2^2 \quad (4.9)$$

Table 4.6: Coefficients in the developed Empirical relationship (Equation 4.9) for the output responses

As Printed (AP)											
	XY build direction						**XZ build direction**				
	ψ	ψ_0	ψ_1	ψ_2	ψ_3	ψ_4	ψ_0	ψ_1	ψ_2	ψ_3	ψ_4
LB	**%el**	26.910	0.142	-0.706	-1.310	-4.461	32.610	-0.659	-0.239	-3.867	-3.146
	YS	640.98	2.546	63.570	-4.259	-136.32	459.59	6.852	-8.262	7.195	32.503
	UTS	1054.07	-18.60	-13.68	14.043	-28.308	975.48	2.459	-28.76	2.392	-23.035
UB	**%el**	41.060	1.570	-1.305	1.310	-2.957	46.290	1.486	-2.365	-1.984	-3.421
	YS	860.27	10.845	9.433	12.473	34.236	857.35	20.878	19.060	39.306	-89.284
	UTS	1224.44	34.289	6.955	13.509	-36.294	1155.30	22.332	-25.99	36.851	5.459
As Printed and solution treatment followed by age hardening (STA)											
	XY build direction						**XZ build direction**				
LB	**%el**	18.100	0.094	-1.264	-0.376	-5.447	16.850	-0.116	-3.685	-0.488	0.589
	YS	1097.92	14.352	25.253	6.795	-194.5	1163.30	54.150	-49.20	-33.48	-124.409
	UTS	1401.73	-13.14	12.288	51.316	-20.68	1434.44	7.976	-44.47	17.017	-95.521
UB	**%el**	28.400	-1.622	-3.337	1.413	1.262	26.410	-0.515	-0.034	-0.488	0.589
	YS	1420.44	-12.33	-64.3	28.619	46.372	1386.49	54.450	-49.20	-33.48	-124.4
	UTS	1649.26	-14.5	-31.34	43.159	-65.25	1519.18	-5.520	-8.272	41.488	-35.483

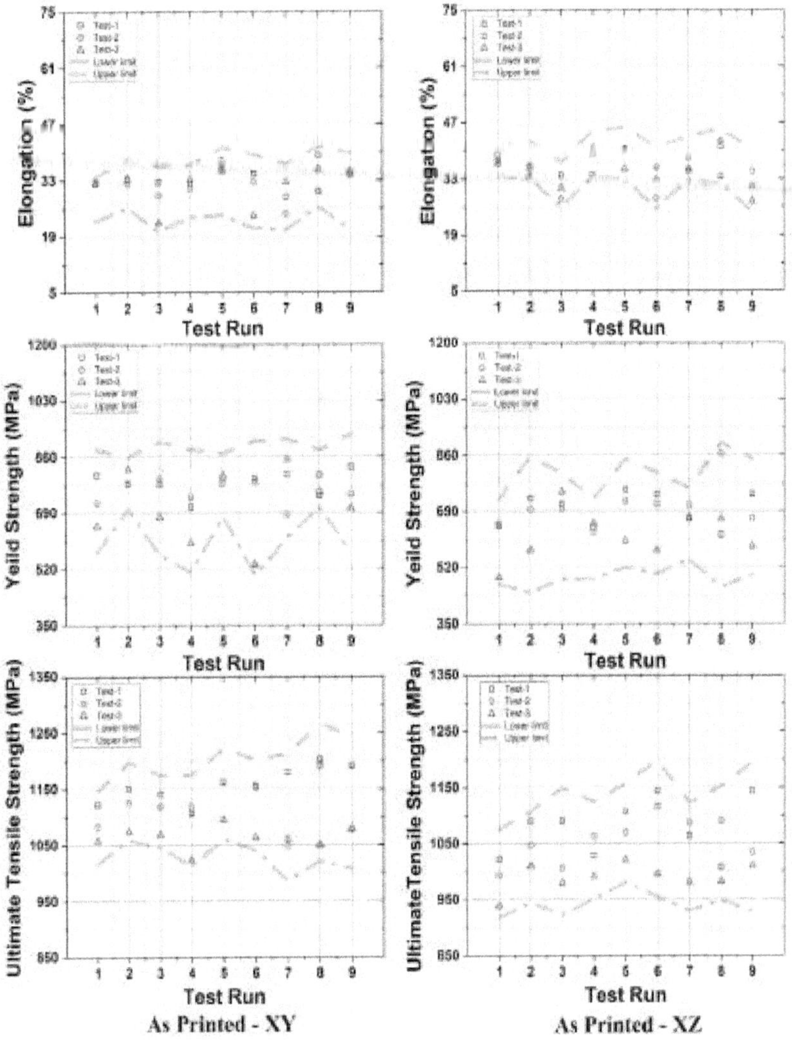

Figure 4.5: Estimations of output responses (XY & XZ – Build directions for sample AP)

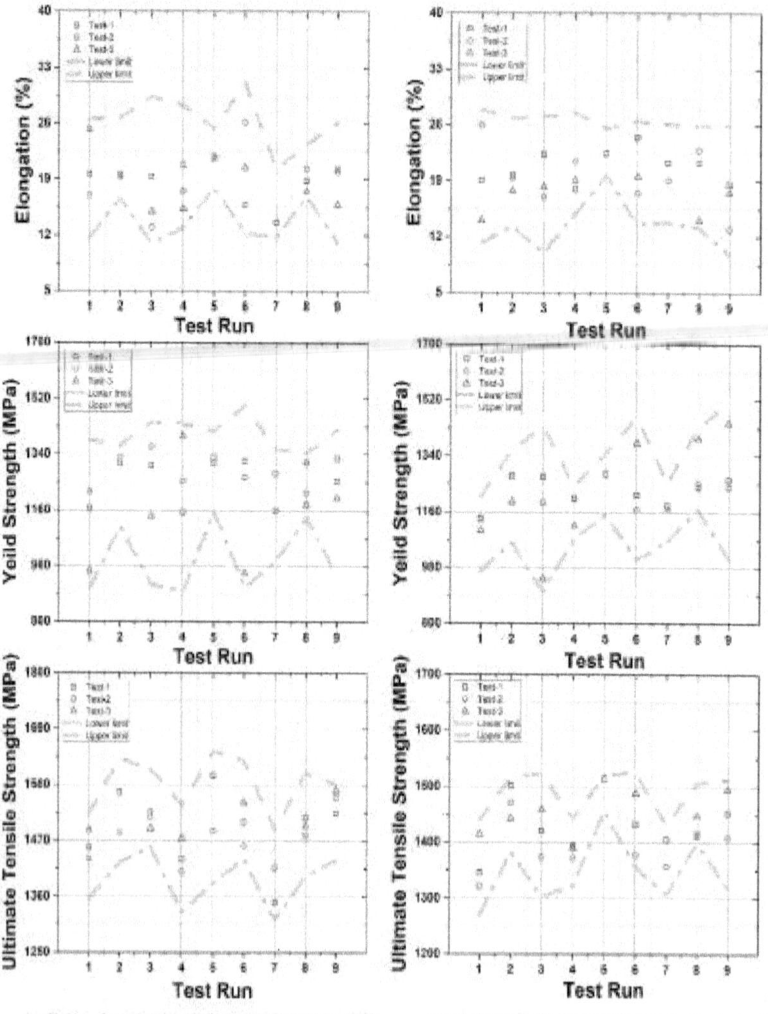

As Printed, Solution Treated and Aged - XY

As Printed, Solution Treated and Aged - XZ

Figure 4.6: Estimations of output responses (XY & XZ – Build directions for sample AP+STA)

Tensile test was conducted on AP and STA and is presented Table-4.6, due to hall patch effect the finer grain size of AM specimen hinders slipping of dislocation thus improving strength and reduction in % elongation as compared to conventionally processed equiaxed IN718 [120]. Additionally, drop in % elongation can be attributed to the presence of porosity, fusion defects, and other microstructural features (such as presence of melt pool and cellular structure) that are typical of additive manufacturing. The samples realized in XY orientation exhibited the higher YS and UTS values with lower % elongation as compared to by XZ orientations. This is due to long columnar grains present in samples realized in XZ direction, which allow for greater amplitude of slip. Similarly, the samples tested at room temperature had higher YS and UTS values than those subjected to 650°C tensile test. By using the empirical relationship, the estimated range was predicted, and test data found to be with in/on the range.

As the strength properties the output responses based on yield strength, optimal parameters were Laser Power, LP (W) = 360 W, Scanning Speed, SS (mm/s) = 960 mm/s, as the Table-4.7 show the corresponding data for the optimum value in both XY, XZ direction along with as printed, as printed with solution treated.

Table 4.7: Range of Mechanical Properties for the optimised parameter

Output responses	AP Range	AP+STA Range
	XY	
YS (MPa)	679 – 870	1156 – 1415
Elongation (%)	24 – 41	18 – 25
UTS (MPa)	1060 – 1221	1386 – 1646
	XZ	
YS (MPa)	520 – 847	1150 – 1349
Elongation (%)	32 – 45	19 – 26
UTS (MPa)	980 – 1157	1455 – 1520

The conventional solution treated and aged IN718 material properties were determined as; Density – 8.19 g/cc, Ultimate tensile strength – 1375 MPa, Yield strength – 1100 MPa, and Elongation – 25 % whereas at elevated temperature (650°C) 1100 MPa, 980 MPa, and 18 % respectively [121]. The microstructural analysis of the SLM IN-718 parts revealed that the print parameters, orientation, and heat treatment conditions significantly affected the porosity and microstructure.

Tensile tested samples with XZ orientation were polished and subjected to optical microscopy in gauge length to check for porosity (see Figure-4.7). The result conform to what was observed with density data. The samples with lowest density showed presence of porosity and higher density showed almost negligible porosity.

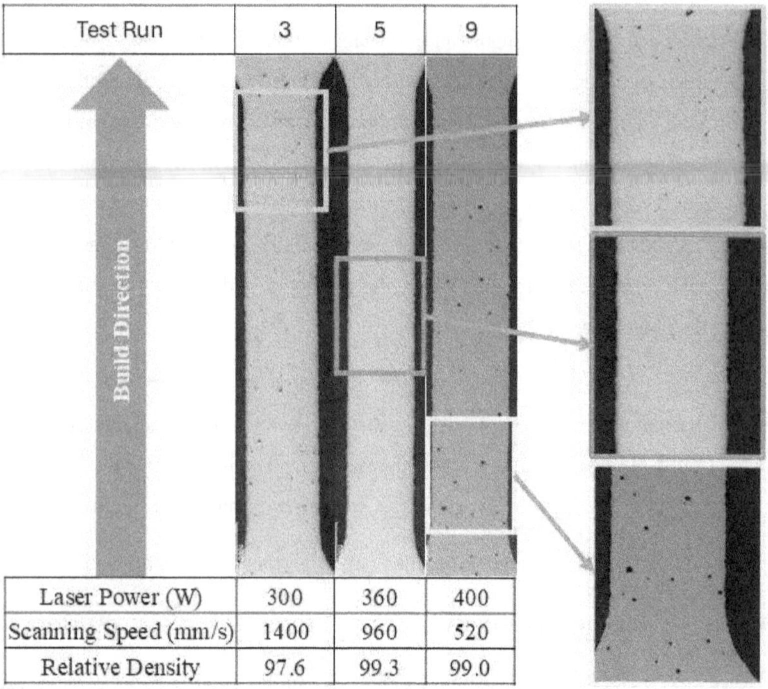

Figure 4.7: Optical microstructures of specific relative density samples with porosities

In AP condition we can see presence of porosity, very fine grains along with well-defined fish scale morphology melt (see Figure-4.8) which is a feature of non-heat treated additive manufactured part. After solution treatment & aging melt pool boundaries have been eliminated while long and thin columnar grain and porosity were still present, which matches with literature [122].

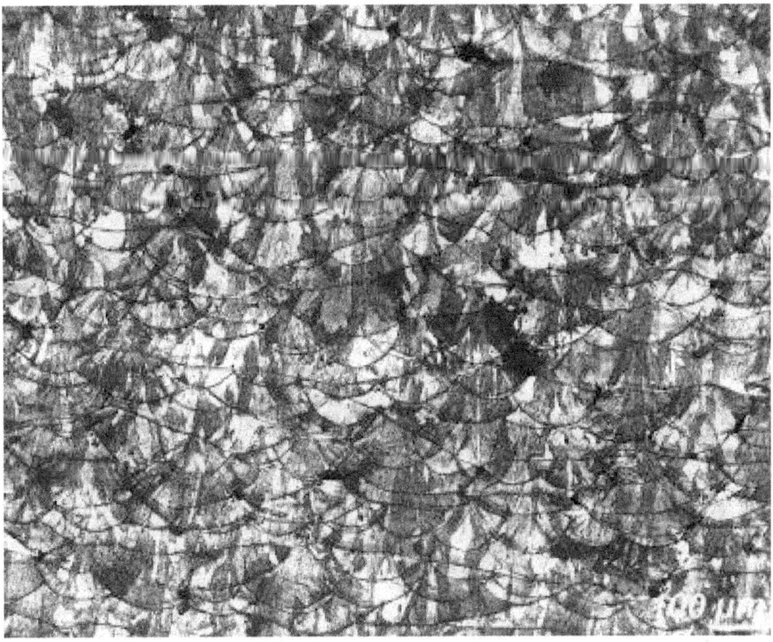

Figure 4.8: As printed sample with no heat treatment

Tensile test was conducted on AP & STA SLM AM IN-718 tensile test specimen which showed that the print parameters, orientation, and heat treatment conditions had a significant effect on the mechanical properties. As printed condition tensile properties YS, UTS & % El are in range of ~800 MPa, 1100 MPa, & 25-30 % in Y direction and ~700 MPa, 1000 MPa,& 30-35 % in Z direction. After heat treatment strength increase by around 500 MPa and elongation reduce by around 50 %. In STA condition tensile properties YS, UTS & % El are in range of ~1200 MPa, 1400 MPa, & 10-20 % in both direction. This increase in strength is due to the presence of precipitation hardening. Table-4.8 shows tensile properties of SLM IN-718 when subjected to tensile testing at 650° C. The Table-4.8 compare properties of Test run number – 7 which have low density (LD = 97.6) and Test run number – 5 which have high density (HD = 99.2) in Y & Z orientation. It is observed that all samples have tensile strength in same range but have variation in elongation. Test run number – 5 being the highly dense had greater elongation in all cases (except for sample STA 7 XY which has elongation around 5 %), Test run number – 7 being the sample with low density showed elongation of ~ 12 % in STA condition.

Table 4.8: Comparison of tensile properties tested at room temperature (RT) and 650° C for STA

Test Run	Build Direction	RT			650° C		
		%el	YS (MPa)	UTS (MPa)	%el	YS (MPa)	UTS (MPa)
7	XY	13	1199	1347	12	1048	1149
	XZ	20	1174	1380	14	944	1080
5	XY	21	1321	1543	5	1071	1175
	XZ	22	1281	1514	[illegible]	1106	1113

The microstructures for the optimum process parameters are shown in Figure 4.9 for the sample with high density and showing negligible presence of porosities (Figure-4.9(a, d)), in both XY, XZ planes. The microstructures in XY plane (Figure-4.9(c)) show equiaxed grains and XZ plane (Figures-4.9(f)) show columnar grains after solution treatment and aging.

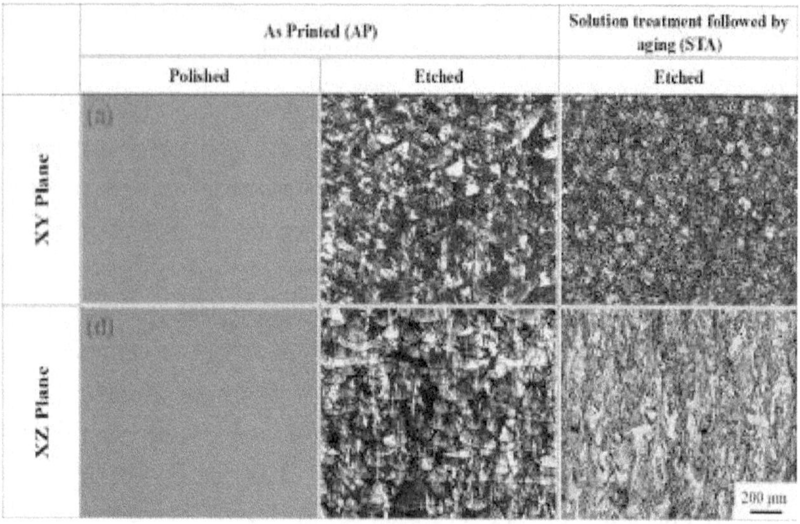

Figure 4.9: Optical microstructures of samples in as printed condition (a,b in XY plane and d, e in XZ plane) and in STA condition (c in XY plane and f in XZ plane), printed with optimum process parameters

Figure-4.10 illustrates the effect of print parameters, print orientation, and heat treatment (HT) conditions on the stress-strain behaviour of IN718. It is observed that the percentage elongation is greater for samples printed in the XZ direction compared to those printed in the XY direction, regardless of the printing parameters and HT conditions. This increase in elongation is due to enhanced pre-necking ductility, while there is no significant change in post-necking ductility. The rise in pre-necking ductility can be attributed to increased work hardening, resulting from heightened dislocation interaction and greater slip activity favoured by long columnar grains aligned with the direction of loading. After STA, both strength and percentage elongation decrease in the XY and XZ directions. Samples printed with optimized parameters (Condition 5 – blue legend) exhibit better elongation across all conditions (Figure-4.10(a) to (d)). In contrast, samples printed with unoptimized parameters (Condition 1, 3, 6, 7 – black, red, aqua, pink legend) show poor elongation due to higher porosity. These findings are consistent with the results of the Taguchi analysis.

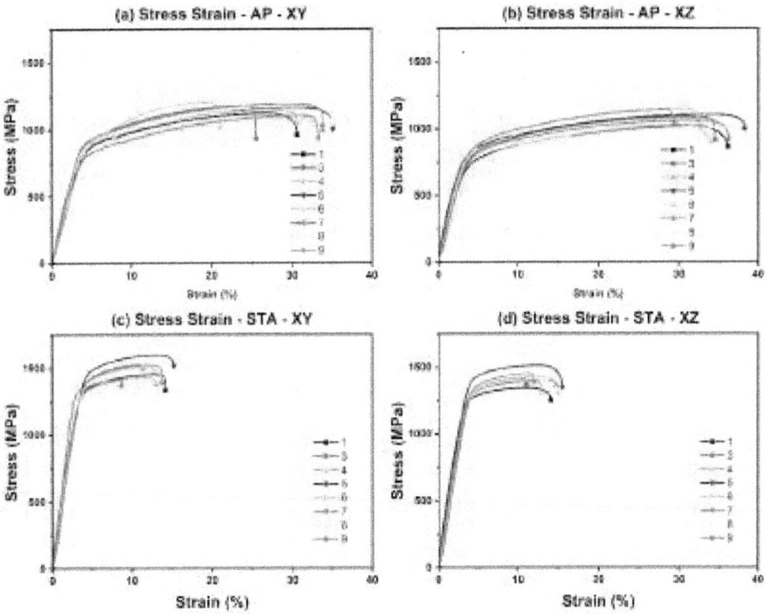

Figure 4.10: Stress Strain Graph of uniaxial tensile test of IN 718 under (a) AP condition – realized in XY direction (b) AP condition – realized in XZ direction (c) STA condition – realized in XY direction (d) STA condition – realized in XZ direction

Estimations are from the developed empirical relation (see Table-4.9) by varying LP from 300 to 400 W: SS from 520 to 1400 mm/s at constant HD of 0.11 mm and PBLT of 0.04 mm can be seen in Figure-4.5 and Figure-4.11. Let ψ be the general function for output response, which can be expressed in terms of input variables as in Equation-4.9 (See Table-4.6 for As printer LB and UB values).

Table 4.9: Coefficients in the developed Empirical relationship (Equation 4.9) for the output responses (HIP)

Hot iso-static pressing (HIP)											
	XY build direction						XZ build direction				
	ψ	ψ_0	ψ_1	ψ_2	ψ_3	ψ_4	ψ_0	ψ_1	ψ_2	ψ_3	ψ_4
LB	%el	19.76	-0.334	-0.739	-0.528	-0.435	24.136	-0.112	-0.167	-1.951	-3.612
	YS	1064.76	67.430	96.118	-65.14	-147.75	1082.43	42.559	70.982	-29.49	-3.307
	UTS	1416.75	4.129	34.126	-43.66	-46.823	1393.26	-5.936	7.714	-7.251	-7.705
UB	%el	22.76	1.815	1.535	-0.370	3.648	29.290	0.538	-1.166	-2.253	-3.499
	YS	1271.17	-21.39	-49.96	1.665	119.891	1256.61	-37.80	-20.22	-49.78	52.331
	UTS	1461.14	7.035	-0.073	-3.402	29.643	1482.46	16.803	1.972	-23.13	0.097

Tensile test was conducted on AP & HIP+STA SLM AM IN-718 tensile test specimens. The tensile testing of the IN-718 parts showed that the print parameters, orientation, and heat treatment conditions had a significant effect on the mechanical properties. Figure-4.5 and Figure-4.11 shows variation of tensile properties of AP & HIP+STA sample with estimated range prediction in both orientation (XY & XZ). Only 9 conditions are shown here since density was measured for 9 conditions only.

In HIP+STA condition the tensile strength decreases by a small amount as compared to STA, this is due to grain recrystallization which occur during Hot isostatic pressing. We observe that when printed at optimum parameters, the HIP treatment is not required as also observed by studies [123].

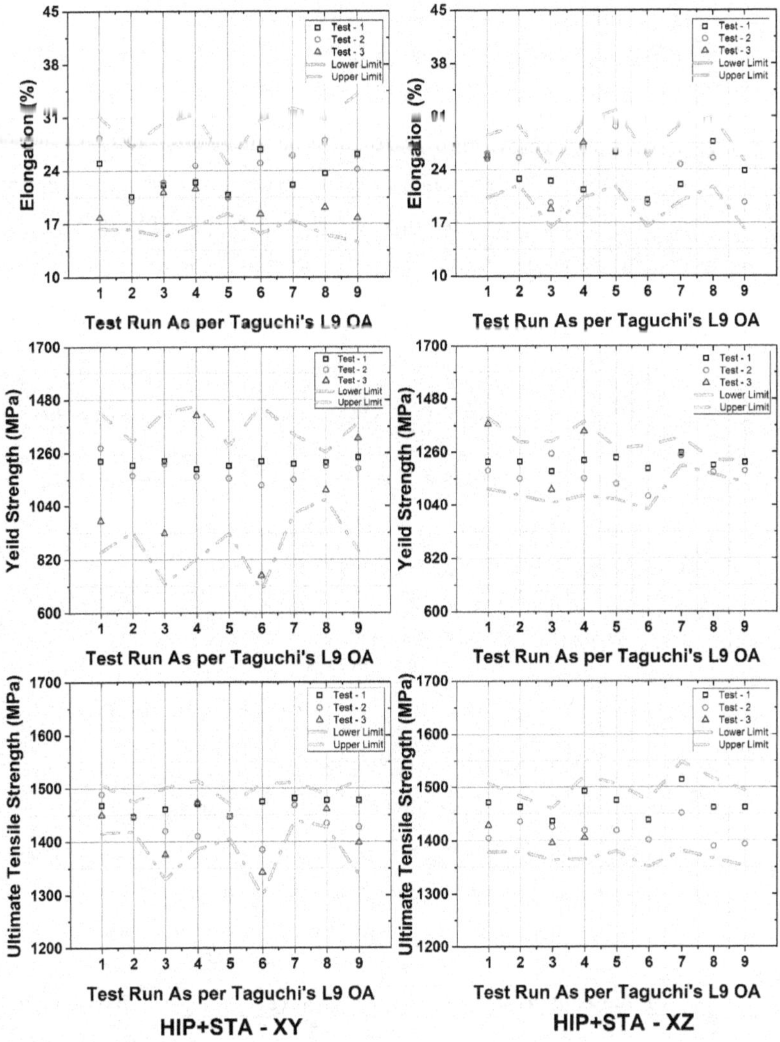

Figure 4.11: Estimations of output responses (XY & XZ – Build directions for sample HIP)

Table-4.10 presents tensile properties of SLM IN-718 when subjected to tensile testing at 650° C. Properties were compared when samples were subjected to AP & HIP+STA heat treatment. It is observed that all samples have tensile strength in same range but have variation in elongation. Test sample No. 5 being the highly dense had greater elongation in all cases (except for sample STA 22 XY which has elongation around 5 %), sample 5 being the sample with low density showed elongation of ~ 12 % in STA condition, even after HIP+STA the elongation did not increase. These properties were compared with room temperature tensile properties in Table-4.10. It was observed that there is a decrease in EL, YS and UTS when tested at 650° C, which is within expectations.

Table 4.10: Comparison of tensile properties tested at RT and 650° C

	Tested at RT			**Tested at 650° C**		
	El	**YS**	**UTS**	**El**	**YS**	**UTS**
XY – 7	24 %/	1182 MPa	1474 MPa	9 %	950 MPa	1090 MPa
XY – 5	20 %	1182 MPa	1447 MPa	17 %	1011 MPa	1112 MPa
XZ – 7	23 %	1250 MPa	1483 MPa	11 %	996 MPa	1120 MPa
XZ – 5	28 %	1182 MPa	1447MPa	25 %	1041 MPa	1140 MPa

As the strength properties the Output responses based on Yield Strength, optimal parameters were Laser Power, LP (W) = 360 W, Scanning Speed, SS (mm/s) = 960 mm/s, as the Table-4.11 show the corresponding data for the optimum value in both XY, XZ direction along with as printed, as printed with solution treated.

Table 4.11: Optimum parameters

Output responses	**AP Range**	**HIP+STA Range**
	XY	
YS (MPa)	679.4 – 869.9	927.1 – 1289.9
Elongation (%)	24.4 – 41.1	18.4 – 24.9
UTS (MPa)	1060.3 – 1221.5	1404.6 – 1471.1
	XZ	
YS (MPa)	520.4 – 847.6	1064.3 – 1276.1
Elongation (%)	32.8 – 45.8	21.8 – 31.7
UTS (MPa)	980.8 – 1157.3	1379.7 – 1506.6

The conventional IN718 material properties were determined as; Density – 8.19 g/cc, Ultimate tensile strength – 1375 MPa, Yield strength – 1100 MPa, and Elongation – 25 % whereas at elevated temperature (620°C) 1100 MPa, 980 MPa, and 18 % respectively [121].

CuCrZr

Figure-4.12(a) displays engineering stress-strain curves of CuCrZr at room temperature and Figure-4.12(b) corresponds to the mechanical response of the samples for different heat treatment conditions. A significant difference in mechanical properties was observed indicating the influence of different heat treatments [124]. The DAH Treatment sample displayed the highest YS and UTS of 410MPa and 438MPa respectively, with the lowest elongation of 8.55%. This increase in mechanical properties is due to the combined effect of high dislocation density and precipitation hardening effect [125] along with coherent FCC precipitates in the copper matrix [126]. The YS, UTS, and ductility of as-printed conditions are 152 MPa, 192 MPa, and 20.2% respectively. The as-printed samples displayed higher ductility potentially attributed to the presence of dislocation networks [127]. These networks allow the bypassing of untangled dislocations leading to enhanced ductility. Abetter improvement in mechanical properties was observed in the STA sample with YS and UTS values of 198 MPa and 245 MPa respectively.

Solution treatment promotes the dissolution of residual stress and dislocations within the matrix due to heat treatment near solutionizing temperatures. This phenomenon resulted in a decrease in strength but concurrently improved ductility, which was observed with an increased ductility value of 12.4%. Surprisingly, the as-printed sample's YS shows a value that is closer to the STA sample. This phenomenon can be attributed to the combined effects of non-equilibrium solidification during rapid cooling and the high dislocation density inherent to metal additive manufacturing (MAM) [128]. Subsequent analysis of samples subjected to brazing cycles indicated YS and UTS values of 176MPa and 232MPa respectively.

However, the brazed samples displayed a significantly higher elongation of 15.5% compared to STA samples with an elongation of 12.4%. The enhanced ductility of the brazed samples is generally attributed to the extended period of exposure to solutionizing temperature and slow cooling which removed the remaining dislocation

[129] and increased the grain growth due to slow cooling enhancing the material plasticity [130].

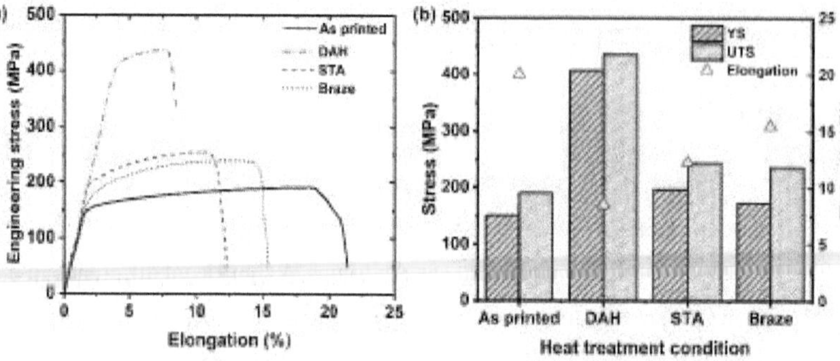

Figure 4.12: (a)Engineering Stress-Strain curves of CuCrZr generated from As printed (ASP), DAH, STA, braze;(b) Mechanical properties variation

Examination of the fracture surfaces in Figure-4.13(a) revealed the distribution of numerous dimples indicating ductile fracture mode. However, variations in dimple size and depth were observed across the heat-treated specimens, which confirm a potential influence of the heat treatment parameters on the material's ductility [55]. The as-printed sample displayed the highest elongation as evidenced in Figure-4.13(b) revealing deeper dimples. In contrast, DAH samples presented with shallow dimples (Figure-4.13(c,d)), characteristic of brittle fracture. Conversely, STA samples also displayed deep dimples (Figure-4.13(e, f)), indicative of enhanced ductility. Gas porosities were observed on the fracture surface due to the dissolution of gas at high temperatures inside the melt pool [131]. Fold morphology was observed along the surface of the dimples indicating strong plastic deformation [128]. Examination of the fracture surface revealed the presence of unfused powder particles. These defects likely originated from either unmelted powder particles trapped within the melt pool or spatter generated during the LPBF process [132].

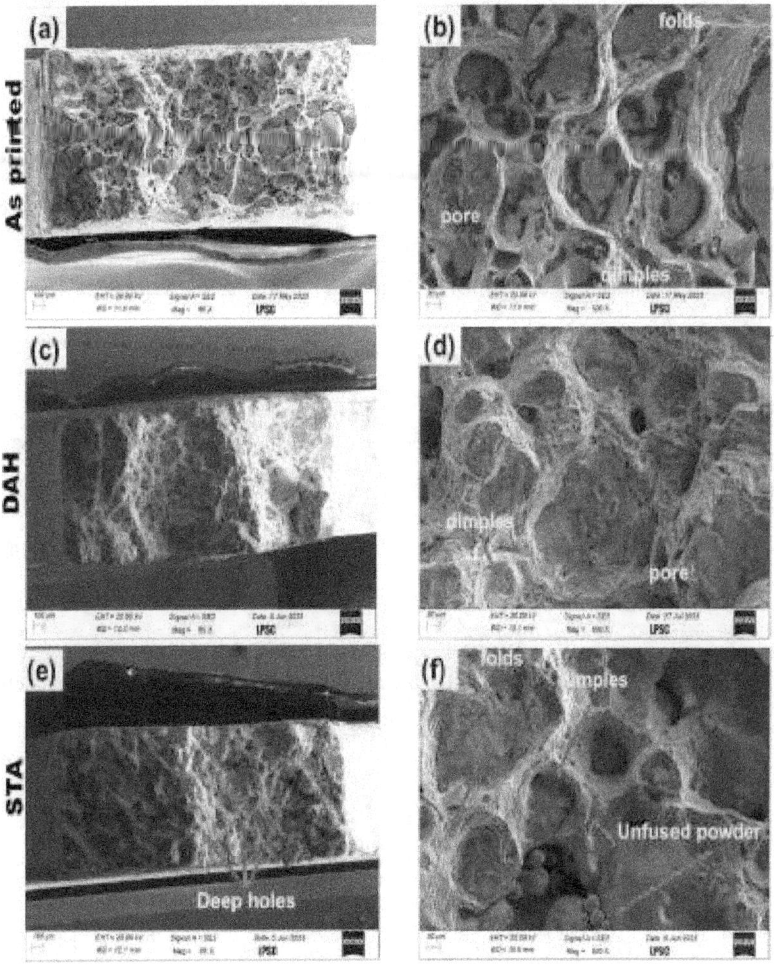

Figure 4.13: Fractography of the CuCrZr tensile tested sample

SS316L

In case of SS316L, in Table-4.12 that percentage contribution of LP, SS, HD and f on the equivalent energy are 34.9 %, 35%, 28.9% and 1.2 % respectively; on R_a are 83.9 %, 4 %, 11.3 % and 0.8 % respectively; and on Hv are 54.8 %, 19.2 %, 21.1 % and 4.9 % respectively. The percentage contribution of the fictitious parameter (f) on the equivalent energy, R_a, and Hv are 1.2 %, 0.8 %, and 4.9 % respectively, which are nothing but the error (%).Empirical relationships (Equations 4.10 – 4.12) are developed for the E_0^*, R_a, and H_v in terms of LP, SS and HD using the additive law from the mean values of ANOVA presented in Table-4.12 in the form:

$$E_0^* = 2.3313 + 0.4095\xi_1 + 0.0305\xi_1^2 - 0.4103\xi_2 + 0.0302\xi_2^2 - 0.3688\xi_3 + 0.0945\xi_3^2 \quad (4.10)$$

$$R_a = 8.0967 - 3.0033\xi_1 + 0.77\xi_1^2 + 0.5567\xi_2 + 0.63\xi_2^2 + 0.4183\xi_3 - 1.785\xi_3^2 \quad (4.11)$$

$$H_V = 215.8744 + 16.3383\xi_1 - 17.6083\xi_1^2 - 10.44\xi_2 - 8.0033\xi_2^2 - 9.385\xi_3 - 12.7783\xi_3^2 \quad (4.12)$$

Here, $\xi_1 = \frac{LP-252}{46}$, $\xi_2 = \frac{SS-1100}{200}$ and $\xi_3 = \frac{HD-90}{15}$.

Figure-4.14(a to c) present the estimates of the performance indicators including test data (Jiang et al., 2019)[67] for all possible 27 parameters.

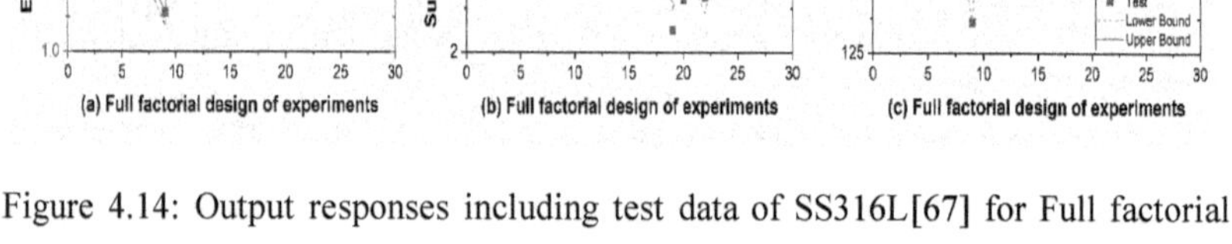

Figure 4.14: Output responses including test data of SS316L[67] for Full factorial design of experiments (a) equivalent energy density, (b) Surface Roughness, (c) Hardness

Table 4.12: Influence of SLM process parameters for SS316L

Parameters	1-Mean	2- Mean	3-Mean	SOS	% Contribution
Equivalent Energy Density (E_0^*); grand mean = 2.436					
LP (W)	2.01	2.42	2.86	1.01	34.9
SS(mm/s)	2.86	2.41	2.04	1.01	35.0
HD (µm)	2.84	2.37	2.10	0.83	28.9
D	2.39	2.40	2.52	0.04	1.2
Surface Roughness, R_a (µm); grand mean = 7.840					
LP (W)	11.10	7.33	5.09	55.31	83.9
SS(mm/s)	7.49	7.42	8.61	2.65	4.0
HD (µm)	6.83	9.03	7.66	7.42	11.3
D	7.90	7.52	8.10	0.52	0.8
Hardness (H_v); grand mean = 190.281					
LP (W)	168.07	202.02	200.75	2221.75	54.8
SS(mm/s)	198.05	195.62	177.17	782.07	19.2
HD (µm)	195.41	198.80	176.64	855.04	21.1
D	191.69	195.20	183.95	198.89	4.9

4.3 Microstructural Observations of Aerospace grade Materials

IN718

The effect of LPBF process parameters on the microstructure and densification of IN718 were examined here in detail. Above 99% densification was achieved for 360 W of LP and the range of SS.

For a SS of 960 mm/s, the melt pool depth has increased, whereas width has decreased with laser power. For a LP of 360W, as the melt pool depth decreased, and width increased relatively with scanning speed. Conduction mode, transition mode and keyhole conduction modes are possible in case of additive manufacturing. Table-4.13 shows the densification of the printed builds and the estimated range with actual test data.

The process parameters, such as LP and SS, play a crucial role in determining the quality and properties of the printed parts. The variation in density observed in this study is attributed to the changes in the LPBF process parameters. By increasing the LP, more energy is delivered to the powder bed, resulting in a deeper melt pool. Conversely, by increasing the SS, the width of the melt pool decreases. Therefore, the combination of higher laser power and lower SS leads to an increased melt pool depth-to-width ratio. Figure-4.15 showcases a montage of optical microstructures of specimens that were printed using different combinations of LP and SS under the experimental conditions studied. These microstructures provide visual evidence of the effect of process parameters on the resulting material density.

Table 4.13: Densification of the printed builds and the estimated range with test data of IN718 for the specified LP (W) and SS (mm/s)

		Scanning Speed, SS (mm/s)						
		300	**520**	**740**	**960**	**1180**	**1400**	
Laser power, LP (W)	**240**	7.618 - 7.727	7.844 - 7.954	7.993 - 8.103	8.065 - 8.174	8.059 - 8.168	7.975 - 8.085	Est. Range
		(7.618)*	(8.107)	(7.995)	(8.035)	(8.109)	(8.05)	Test Data
	280	7.668 - 7.777	7.895 - 8.004	8.043 - 8.153	8.115 - 8.224	8.109 - 8.218	8.026 - 8.135	Est. Range
		(7.613)	(8.096)	(7.993)	(8.096)	(8.115)	(8.153)	Test Data
	360	7.690 - 7.800	7.917 - 8.026	8.066 - 8.175	8.137 - 8.247	8.131 - 8.241	8.048 - 8.157	Est. Range
		(7.797)	(7.983)	(8.193)	(8.141)	(8.108)	(8.128)	Test Data
	400	7.663 - 7.772	7.889 - 7.998	8.038 - 8.147	8.109 - 8.219	8.103 - 8.213	8.020 - 8.129	Est. Range
		(7.772)	(8.024)	(8.138)	(8.165)	(8.118)	(8.08)	Test Data

The VED in Equation (1) consists of the LPBF process parameters (LP, SS, HD and PBLT). For any specific VED value, it is possible to create many sets of the LPBF process parameters. An attempt is made to obtain the densification for the created sets of the LPBF process parameters to examine the influence of VED. Table-4.14 and Table-4.15 give the achieved densification for the created sets of LPBF process parameters for which VED = 67.5 J/mm^3. It is clearly visible that for a constant value of VED = 67.5 J/mm^3, appreciable variation in the densification is observed. This variation is not due to the experimental scatter. It is mainly due to the variation in the sets of the LPBF process parameters. Hence the densification cannot be represented explicitly in terms of VED. Since, there are 4 LPBF process parameters, two case studies were made. In the first case, two parameters (LP and SS) are varied by keeping HD and PBLT at constant. In the second case, two parameters (LP and HD) are varied by keeping SS and PBLT constant.

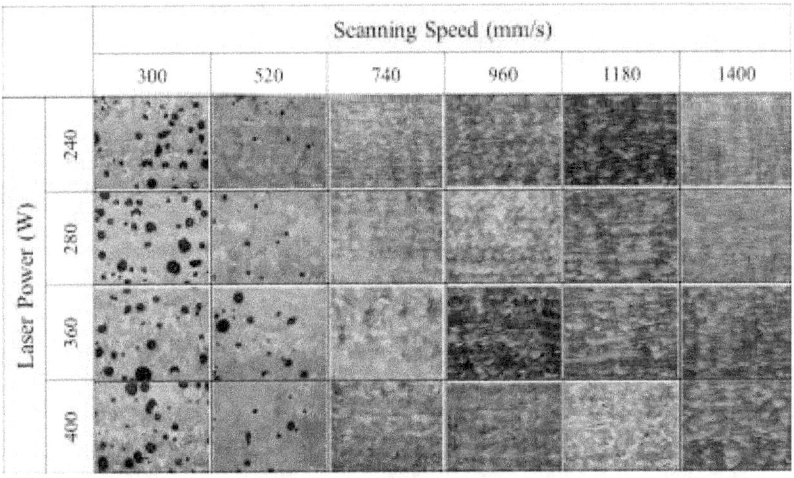

Figure 4.15: Montage of optical microstructures of specimens printed using various scan power and scanning speed at all experimental conditions studied

Table-4.14: Density for a constant VED, HD of 0.11 mm and LT of 0.04 mm

LPBF process parameters			**Density (g/cc)**	
LP (W)	**SS (mm/s)**	**VED (J/mm^3)**	**Test**	**Estimated Range**
240	808	67.5	8.103	8.024 - 8.133
260	876	67.5	8.12	8.075 - 8.184
280	943	67.5	8.16	8.112 - 8.221
300	1011	67.5	8.108	8.136 - 8.245
320	1078	67.5	8.101	8.145 - 8.255
340	1145	67.5	7.763	8.142 - 8.251
360	1213	67.5	8.062	8.124 - 8.233
380	1280	67.5	8.097	8.092 - 8.202
400	1347	67.5	8.136	8.047 - 8.156

To further understand the relationship between process parameters and build quality, Table-4.15 presents the densification results of specimens at a constant VED, while varying the laser power and hatch distance. Table-4.15 highlights how the process parameters impact the achieved density of the 3D printed parts. It emphasizes the dependency of build quality on these specific process parameters, indicating that adjusting laser power and hatch distance can lead to variations in densification. It emphasizes the significance of optimizing laser power, scanning speed, and hatch distance to ensure consistent and controlled fabrication of defect-free parts.

Table 4.15: Density for a constant VED and PBLT of 0.04 mm

LPBF process parameters			**VED (J/mm^3)**	**Density (g/cc)**	
LP (W)	**SS (mm/s)**	**HD (mm)**		**Test**	**Range**
240	960	0.0926	67.5	8.15	8.053 - 8.163
260	960	0.1	67.5	8.103	8.052 - 8.162
280	960	0.1081	67.5	8.1	8.049 - 8.158
300	960	0.1158	67.5	8.012	8.042 - 8.151
320	960	0.1235	67.5	8.161	8.032 - 8.141
340	960	0.1312	67.5	8.116	8.019 - 8.128
360	960	0.139	67.5	8.145	8.003 - 8.112
380	960	0.1467	67.5	8.041	7.984 - 8.093
400	960	0.1544	67.5	8.123	7.962 - 8.071

Although energy density represents the energy input to the process, it should no longer be considered a process parameter. It can have a value for multiple combinations of input process variables with different performance characteristics. This aspect is demonstrated in this paper with test data. Modified Taguchi's method has been followed to develop empirical relationships to the performance characteristics in terms of process variables and demonstrated their validity through comparison of test data. The method suggests few tests as per the orthogonal array and provides complete information for all combinations of levels and process variables. The method also provides the estimated range of output responses so that the scatter in the repeated tests can be assessed prior to the tests.

The LPBF process has a significant advantage in terms of producing complex parts with high material utilization. However, it is influenced by several factors, including laser power, scanning speed, hatch distance, scan strategy, powder material, powder size, and morphology [133], absorption and transmission of laser energy, rapid melting and solidification of material, microstructure evolution, and materials evaporation [134], which are all part of the LPBF process. The presence of microscopic defects can affect the mechanical properties of the fabricated part [135–137].

Lack-of-fusion (LOF) refers to a defect where there is incomplete fusion between the base layer and new layer. This can result in compromising the structural integrity of the component. As shown in optical photomicrographs, the quality of printed parts can be assessed from various types of defects, including LOF. In Figure-4.16(a), the optical micrograph shows a horizontal lack of fusion (H-LOF). This indicates that there is incomplete fusion along the horizontal plane between the adjacent areas. The image displays a visible region where fusion is not adequately achieved. The H-LOF defect can be problematic in applications where the part is subjected to significant loads. In Figure-4.16(b), the optical micrograph shows a vertical lack of fusion (V-LOF). This suggests that there is incomplete fusion along the built plane between the base layer and new layer. Similar to H-LOF, V-LOF can result in compromised structural integrity. The image reveals a lack of bonding/gap in the fusion region along the vertical axis.

By visualizing these optical micrographs, quality of builds can be assessed from the types of defects present. This information is valuable for ensuring the reliability and performance of the printed parts. Corrective actions can then be taken to rectify the LOF issues, by adjusting process parameters.

The LOF defects form because the metal powders are not fully melted enough to deposit a new layer on top of the previous one.Figure-4.17(a) shows the LOF and Figure-4.17(b) shows LOF scanning track in XY Plane. LOF may be considered as (i) poor bonding defects caused by a lack of molten metal during the solidification process, as shown in Figure-4.18(a) and (ii) defects caused by un-melted metal powders, as shown in Figure-4.18(b). When the laser energy is insufficient to cause sufficient molten pool penetration, interlayer bonding defects form. As a result, LOF defects are commonly found in the scan tracks and deposited layers. In a continuous deposition process, interlayer defects may gradually extend and propagate upwards to form large multi-layer defects.

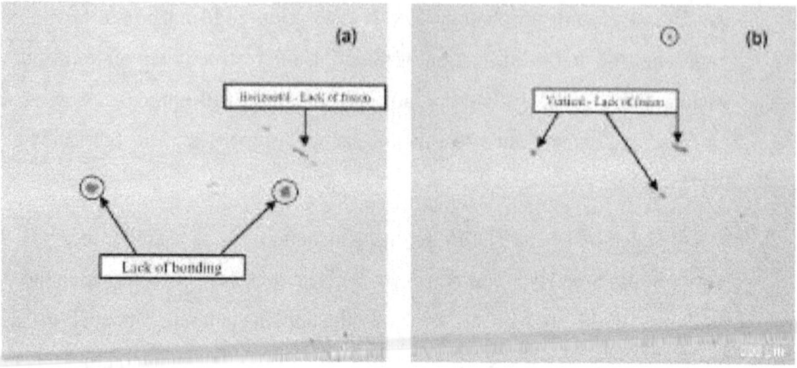

Figure 4.16:OM of samples showing (a) H-LOF in XY Plane and (b) V-LOF in XZ Plane. Circles in both (a) and (b) show lack of bonding between powder particles

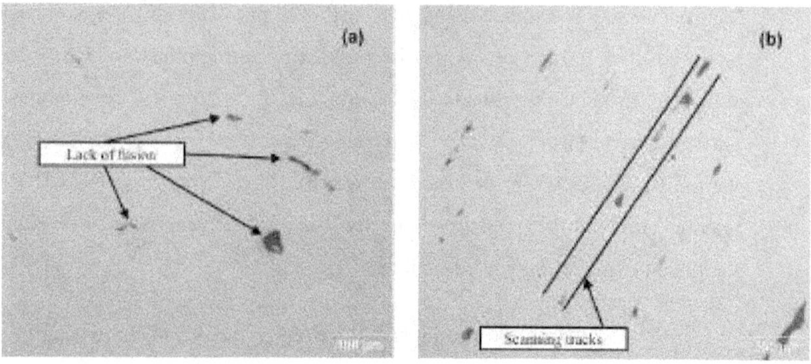

Figure 4.17: OM of samples showing (a) LOF in XY Plane and (b) LOF in scanning track

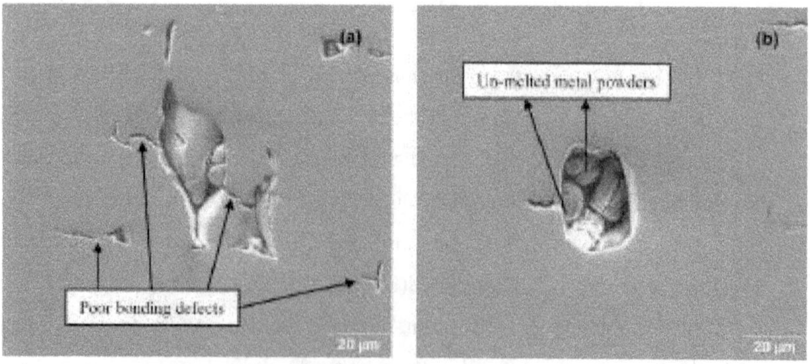

Figure 4.18: SEM of LOF defects (a) due to poor bonding and (b) un-melted metal powders

The optical micrograph in Figure-4.19(a) shows LOF due to deficiency in remelting of melt pools and Figure-4.19(b) shows deficiency in fusing of scanning tracks. Crack initiation and propagation in fabricated parts are caused by a high temperature gradient combined with large residual stresses [138,139].Figure-4.20(a) and Figure-4.20(b) show the formation of inter-layer cracks with delamination. Intra-layer cracks are formed in a layer during solidification of the melt pool between the solidified layer and the printed layer (Figure-4.21(a)).Shrinkage and residual stresses in the melt pool causes intra-layer cracks (Figure-4.21(b)). Intra-layer cracks have a shorter length because they typically form within the melt pool's dimensions.

The density, range and test data for the samples identified with significant defects is presented in Table-4.16. The introduction of undesirable thermal residual stresses, LOF porosities during the process are unavoidable. Some experimental trails were made to correlate the porosity fraction with the melt pool geometry and LPBF process parameters [140–142]. However, the influence of porosities on density of the realised parts need special attention for quantification.

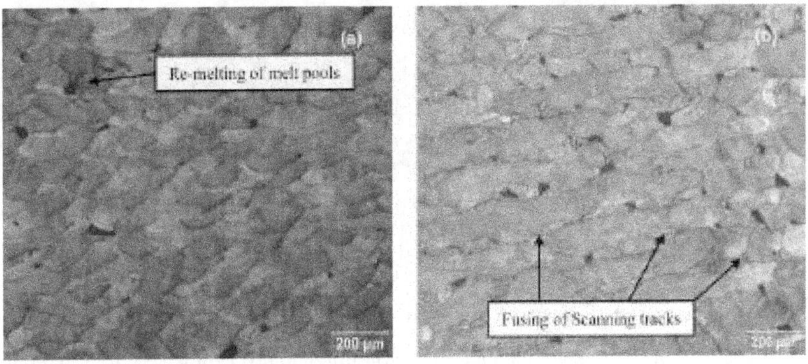

Figure 4.19: OM images showing LOF defects formed by deficiency in (a) remelting of melt pools and (b) fusing of scanning tracks

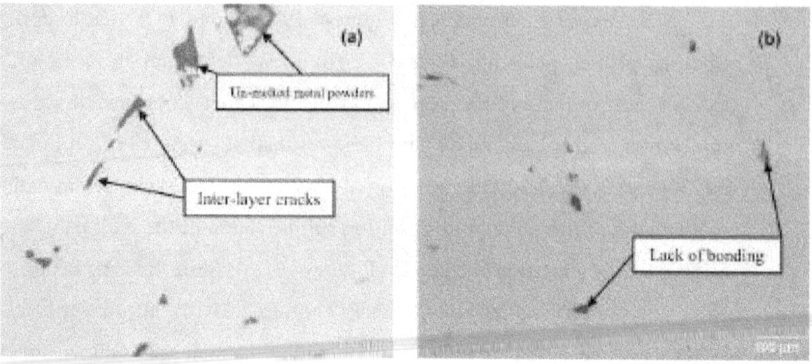

Figure 4.20: OM images of inter-layer cracks (a) between two fusion layers and (b) lack of bonding

Figure 4.21: SEM images of intralayer cracks (a) between solidified layer and the layer being printed and (b) un-melted metal powder

Table 4.16: Defects identified in samples at different LPBF process parameters (constant hatch distance, HD=0.11 mm and layer thickness, LT = 0.04 mm)

LPBF process parameters			**Density (g/cc)**		**Defects**
LP (W)	**SS (mm/s)**	**VED (J/mm^3)**	**Test**	**Range**	
400	300	303.03	7.772	7.663 – 7.772	Porosity
240	1400	38.96	8.050	7.975 – 8.085	Lack of Fusion
400	1400	64.93	8.080	8.020 – 8.129	Cracks

The LPBF process parameters (viz., LP, SS and HD) were identified for IN718 powder to achieve maximum density (ρ). It may be noted that for each value of VED, it is possible to have multiple combinations of input process parameters. Hence different performance indicators can be expected for each combination of input process parameters. This implies that different values of the performance indicators for each VED are not due to experimental scatter. Modified Taguchi approach has been adopted by considering the data of few tests and data has been generated for the full factorial design of experiments. ANOVA has been performed to assess the significance of the LPBF process parameters on density. Test results are found to be close to the expected range of density. The developed empirical relations can be used for estimating the performance responses to the specific LPBF process parameters.

A data-driven model has been proposed to predict the densification behaviour of LPBF IN718 alloy for process parameters; LP, SS and HD. The SS is the most influential process parameter, whereas hatch distance is ranked lowest in terms of percentage contribution. Detailed analysis of defects noticed in metallographically polished and etched specimens has helped in identifying the formation mechanisms of commonly seen defects in LPBF IN718. This study has brought out a complete understanding on the densification and the effect of LPBF process parameters on the generation of microscopic defects. Defects being a major disadvantage of AM processed parts, studies on their formation have brought significant clarity in selecting process parameters.

Figure-4.22(a) and (b) shows the hardness contour plots with laser power and scan speed. The maximum hardness is at 280W laser power and 960mm/s scan speed. In XZ and XY planes (Figure-4.22) at 320W laser power (LP), the hardness increases with scanning speed (SS) up to 960 mm/s and then after decreases.

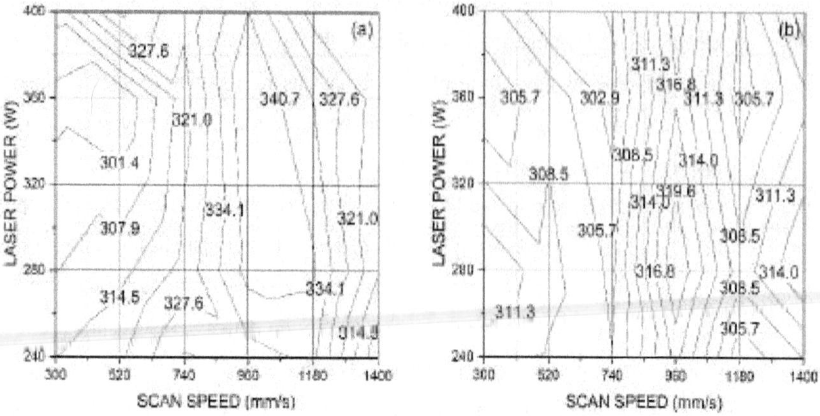

Figure 4.22: Contour plot showing variation of hardness in (a) XZ plane and (b) XY plane

Hardness can be defined as the resistance of a material to localised plastic deformation. In order to ensure reliability and repeatability of mechanical properties the 3D printed parts, process parameters must be optimized. According to Babu et al. [68,143] directional cooling results in anisotropic mechanical properties in AM products. The hardness of sample in as printed condition need not be uniform due to the effect of multiple process parameters involved in additive manufacturing such as laser scan power, scan speed and hatch distance. However, on solution treatment followed by artificial aging (as per AMS 5662 specification for IN718), uniformity can be improved in mechanical properties.

Figure-4.23 the changes observed in microstructure with different heat treatment. In AP condition we can see presence of porosity, very fine grains along with well-defined fish scale morphology melt which is a feature of non-heat treated additive manufactured part. After solution treatment & aging we see that melt pool boundaries have been eliminated while long and thin columnar grain and porosity were still present, which matches with literature [57]. After HIP+STA treatment we see that porosity has been eliminated and long columnar grains have been replaced with homogenous recrystallized equiaxed grains.

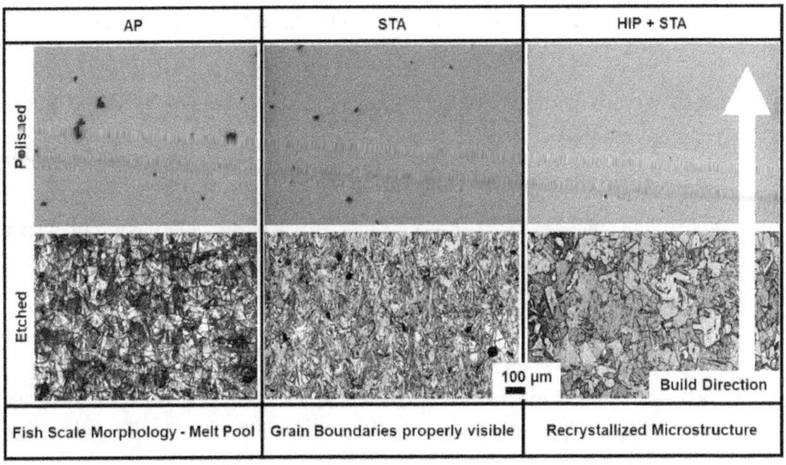

Figure 4.23: Effect of Heat Treatment on Microstructure

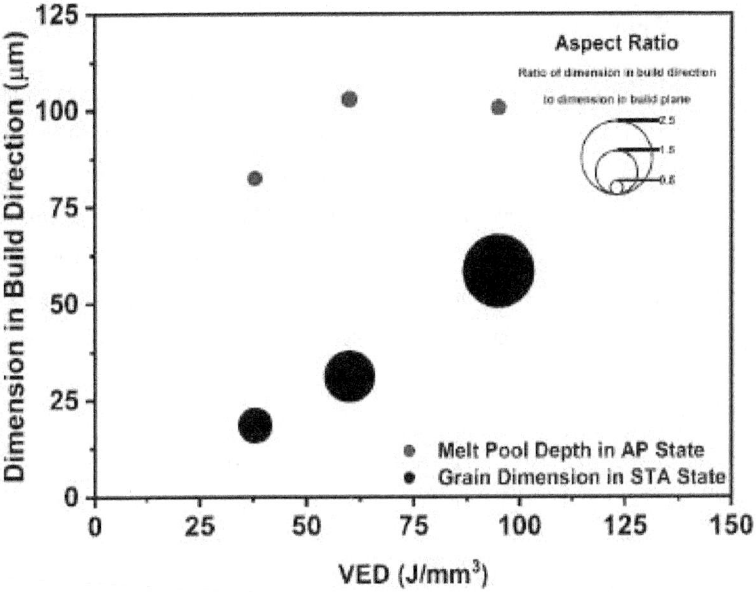

Figure 4.24: Effect of VED on melt pool and grain sizes of specimens in AP and STA conditions respectively

The effect of VED on the dimensions of the melt pool and grain size was investigated for the samples in AP and STA conditions, and the same is illustrated in Figure-4.24. The melt pool depth (and aspect ratio) for samples printed with VED of

38, 58, and 92 J/mm3 were observed to be 82 (0.5), 103 (0.55), and 101 (0.51) μm, respectively. Melt pool depth increased from VED of 38 to 58 J/mm3, then plateaued till 92 J/mm^3. No appreciable change in the aspect ratio of melt pool was observed. This suggests that there is an upper limit to the penetrability of the melt pool, even at higher VED. This provides an approximation of optimum parameters for printing, as any further increase in VED any further will lead to vaporization of the melt pool, creating keyhole defects. However, there a different observation with grain size. It was observed that grain size in build plane continued to increase without any sign of plateauing. The grain size in build plane (and aspect ratio) for samples printed with VED of 38, 58, and 92 J/mm3 were observed to be 18 (1.1), 31 (1.7), and 58 (2.4) μm, respectively. Along with grain size, the aspect ratio of columnar grains continued to increase. Further investigation in grain growth kinetics is required to explain this behaviour.

CuCrZr

The morphology of the powder particles generated through gas atomization is depicted in Figure-2.3(b). It is noted that the powder particles exhibit a predominantly spherical morphology with a limited presence of satellite particles. The powder exhibited a particle size range of 12–48 μm as illustrated in Figure-2.3(c) (D_{10} = 15.02 μm, D_{50} = 29.99 μm, D_{90} = 42.3 μm). The X-ray diffraction (XRD) analysis revealed prominent peaks corresponding to the (111), (220), and (200) crystal planes of copper, without any indication of supplementary phases. A comparable finding was observed for the samples produced via additive manufacturing, along both the XZ and XY planes. The exceedingly rapid cooling rates associated with metal additive manufacturing, which can reach up to 10^6 K/s [144], inhibit the diffusion of secondary phases from the primary matrix. This phenomenon results in the establishment of a supersaturated solid solution, as evidenced by the pronounced copper peaks present in both the powder and as-printed samples.

The microstructures of the samples corresponding to different relative densities, in as-printed and polished conditions are shown in Figure-4.25. Image analysis software (ImageJ) was used to estimate the relative density of the samples from the OM images, which closely match with the measured relative densities obtained from Archimedes principle, as presented in Table-4.17.

Table 4.17: Comparison of relative densities

Method	Relative Density (%) (Figure-4.25)					
	a	b	c	d	e	f
Archimedes principle	99.57	99.61	99.88	91.61	94.47	96.19
ImageJ	99.47	99.59	99.86	91.63	95.06	96.11

Porosity in SLM arises from three primary factors; the first being lack of fusion, where the melt pool fails to adequately overlap with adjacent layers or pools [145]. Such pores exhibit a relatively large and non-circular morphology with sharp edges as illustrated in Figure-4.25. These defects are typically associated with low VED conditions. Additionally, lack of fusion pores can emerge from powder denudation. High laser power combined with lower scan speeds can displace surrounding powder upon laser contact, leading to insufficient powder availability during multi-track scanning, which results in lack of fusion pores [146] as illustrated in Figure 4.25.

The second type of pores that form during SLM are gas pores. These pores arise due to the entrapment of inert gases used for oxidation protection or due to dissolved gases within the melt pool. These pores typically exhibit a spherical shape.

Balling phenomenon is another type of defect formation from a LOF which results in unstable and discontinuous melting tracks [147–150]. Balling defects are generally classified into ellipsoidal (diameter up to 500μm) and spherical balling defects(diameter up to 10μm). Presence of ellipsoidal balling defects can be detrimental due to the fact that they tend to introduce coarse surface along with blocking powder paving process which occurs when ball diameter is larger than layer thickness [151] which further results in formation of lack of fusion pores and cracking surrounding the balls which is clearly presented in Figure-4.26. The specimens in Figures-4.26(a) and 4.26(b) exhibit a relative density of 97.75%, while those in Figures 4.26(c) and 4.26(d) show a relative density of 95.42%.

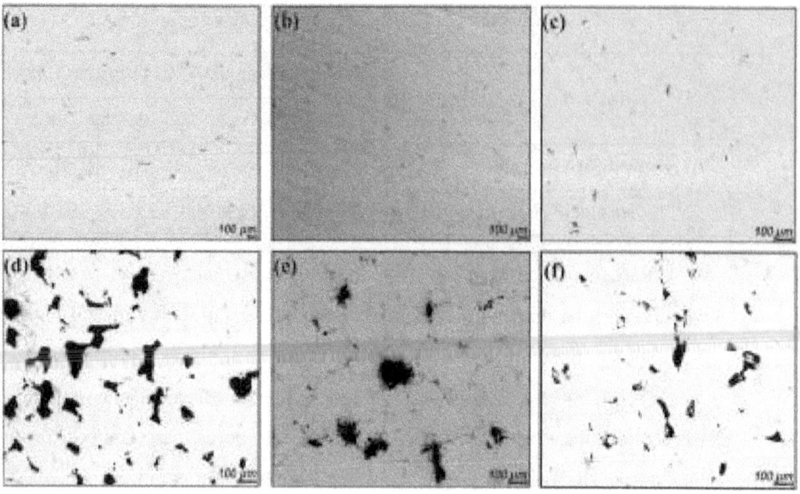

Figure 4.25: Optical micrographs of the as-printed samples in the as-polished condition; whose relative densities (obtained by using Archimedes principle) are: (a) 99.57%; (b) 99.61%; (c) 99.88%; (d) 91.61%; (e) 94.47% and (f) 96.19% respectively.

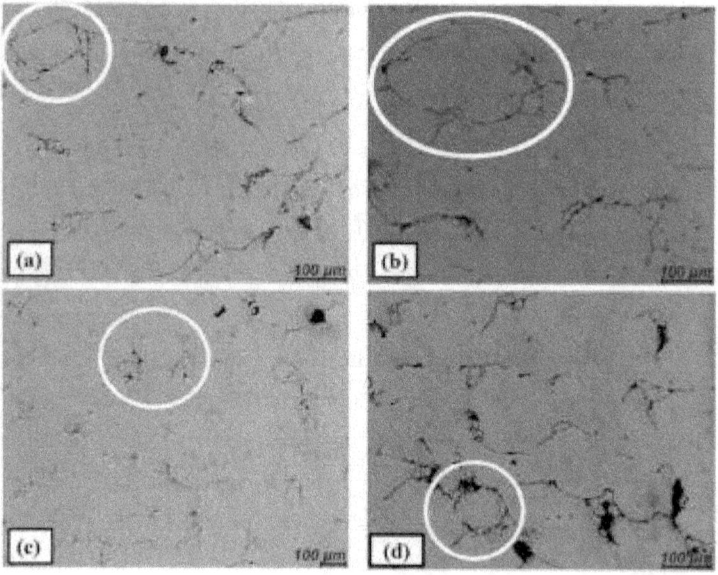

Figure 4.26: Balling defects observed in the optical photomicrographs of specimens printed (a,b) for LP=370 W, SS=610mm/s, HD=0.18 mm, and PBLT=60 μm; (c,d) LP=345 W,SS= 610mm/s, HD=0.18mm, and PBLT=45 μm. The defects are shown in solid circles

Two distinct categories of cracks are typically identified in the SLM process. Due to the substantial localized energy input and the rapid cooling processes, metallic powders are subjected to swift melting and subsequent solidification, which generates considerable internal stresses within the molten pool and among the surrounding melt pools. Inter-layer fractures manifest when the residual compressive stress surpasses the adhesive strength between the surfaces of the molten pools[68].

The significant difference in the freezing points of the alloying elements leads to segregation within the interdendritic region during the solidification process in the melt pool. Hence solidification cracks form above the solidification temperature and propagate along the interdendritic region [152]. Figure-4.27 presents the location of different types of cracks with respect to the melt pool. Intralayer cracks were commonly found along the layer-by-layer melt pool boundaries, while solidification cracks were commonly found within the melt pool, propagating along track-by-track melt pool boundaries. Figure-4.27(c) represents a solidification crack with its typical dendritic head observed inside the crack.

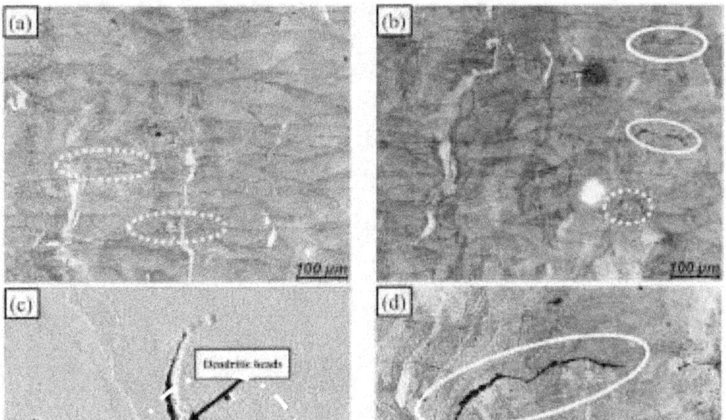

Figure 4.27: Microstructures showing interlayer cracks and intralayer cracks with respect to melt pool boundaries for specimens printed for LP is 370W, SS is 610mm/s, HD is 0.1mm, and PBLT is 30 μm. Intralayer cracks are marked as solid lines and solidification cracks are shown as dotted lines

The grains located in proximity to the melt pool exhibit orientation that is perpendicular to the direction of the laser scanning (XY plane), whereas the grains situated nearer to the boundaries of the melt pool align at an angle as a result of the reheating of the previously deposited material. This phenomenon culminates in a stratified microstructure characterized by coarse (elongated) and fine (short) grains, which is distinctly illustrated in Figure-4.28(d-f). The heterogeneous microstructural properties of grains can be explained through the foundations of solidification theory, where the temperature gradient (G) and the growth rate (R) determine the morphological traits and dimensions of the grains [153]. Due to higher thermal gradient at the boundary of the melt pool, solidification takes place spontaneously via epitaxial growth from partially melted grains or the substrate which results in formation of columnar grains due to low growth rate and high thermal gradient and equiaxed dendrite forms as the solidification progresses to the melt pool due to high growth rate and low temperature gradient [154].

Figure 4.28: The optical microstructures of the as-printed samples in the polished and etched condition; (a-c) represent samples with density values of 99.57%, 99.61%, and 99.88%, respectively for samples printed in XY plane and (d-f) for samples printed in XZ plane

Throughout the process of solidification, the grains show a tendency to develop in a direction that is perpendicular to the melt pool boundaries, directed towards the region of peak thermal flux (along Z direction). For columnar grains, the growth facilitates along the easiest growth direction which is <100> for most FCC structure. Indicating grains with <100> prefer to grow quickly than the other grains resulting in a heterogeneous structure. This phenomenon is generally known as competitive grain growth which is commonly observed in FCC alloys[155]. While on the other hand due to the direct interaction with the laser along XY plane, equiaxed grains were observed along the centre of the laser track (low G/R) and columnar grains observed at the periphery of the laser track (high G/R).

SS316L

The microstructure analysis of SS316L reveals that the density trends initially increase with rising normalized equivalent energy density (E_0^*) but slightly decrease when E_0^* exceeds approximately 2.80 [67]. At low E_0^* values, significant porosity appears, including large irregular pores and smaller spherical ones (Figure-4.29(a), (b)). At higher energy densities, vertical cracks spanning multiple melt pools are observed (Figure-4.29(c), (d)), likely due to deeper penetration, keyhole effects, and higher thermal gradients resulting from slower scan speeds. Additionally, incomplete melting, particularly between adjacent melt pool boundaries, is also observed (Figure-4.29(e), (f)). These findings align with Cherry et al. [50], who found minimum porosity at an energy density of 104.52 J/mm³, equivalent to a normalized energy density of 2.81 in this study.

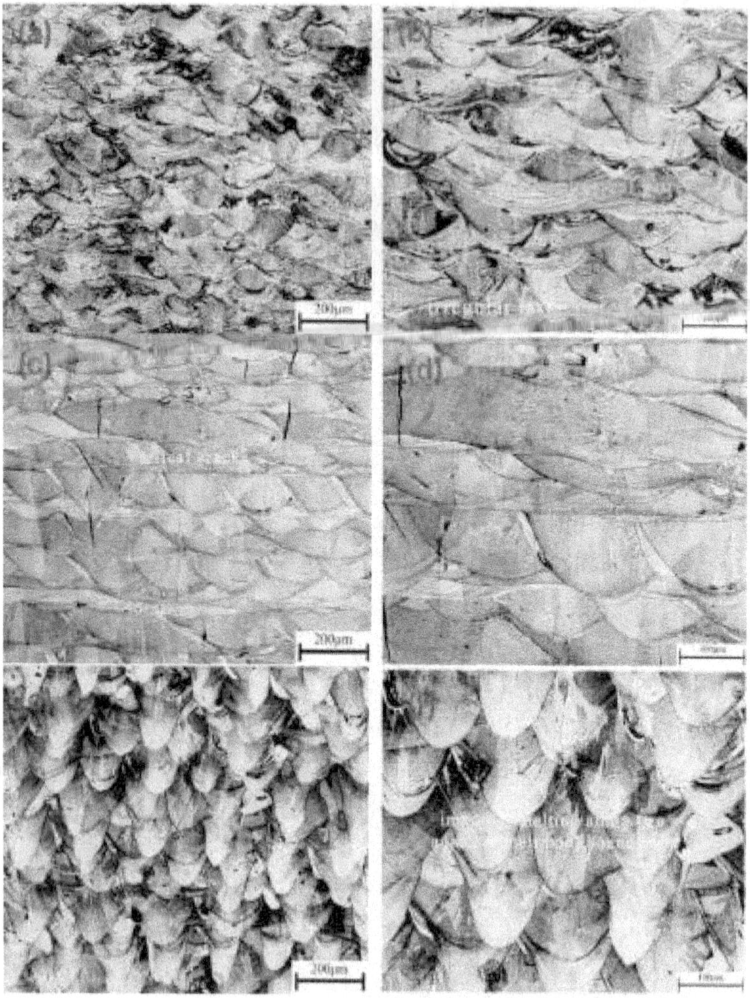

Figure 4.29: OM micrographs showing pores, vertical cracks and improper melting [67]

4.4 Summary of Chapter-4

This chapter presents the measured data from samples of aerospace-grade materials (such as SS316L, IN718, and CuCrZr) through different SLM process conditions as per the Taguchi's L_9 OA and examined the densification behavior, mechanical properties and microstructural analysis.

CHAPTER – 5

Summary of Findings and Future Scope

5.1 Novelty

The novelty of this study is the exploration of optimizing LPBF process parameters for aerospace-grade materials, emphasizing SS316L, IN718, and CuCrZr. Using a modified Taguchi methodology, it establishes empirical relationships between LPBF process parameters (laser power, scanning speed, hatch distance, and powder bed layer thickness) and performance indicators (such as relative density and mechanical properties). Unlike conventional studies that rely on volumetric energy density (VED) as a primary metric, this research shows that VED alone does not capture the complexities of defect formation and material performance. By highlighting the limitations of VED and introducing optimized conditions fitted to each material, this study offers a more significant approach to achieving maximum densification and enhanced mechanical properties, specifically for aerospace applications.

5.2 Investigations

The optimization of Laser Powder Bed Fusion (LPBF) process parameters is important for achieving maximum densification and enhanced mechanical properties in parts across different aerospace-grade materials. This research employed a modified Taguchi approach to optimize key LPBF parameters, including laser power (LP), scanning speed (SS), hatch distance (HD), and power bed layer thickness (PBLT), for SS316L, IN718, and CuCrZr alloys.

For **SS316L**, the study concluded that the equivalent energy density (E_0^*) can be directly evaluated from the input process parameters, though it should not be considered a dependent process parameter due to its variability in producing multiple results. The optimized process parameters were laser power, LP = 298 W, scanning speed, SS = 900 mm/s, and hatch distance, HD = 0.075 mm, resulting in surface roughness, R_a = 2.77μm, relative density, RD = 99.24%, and hardness, H_v = 334 Hv.

In the case of **IN718**, optimization of LPBF parameters led to the identification of optimal conditions: LP = 320 W, SS = 850 mm/s, and HD = 0.11 mm, which resulted in a hardness of 334 Hv. Defects such as lack of fusion (LOF) and un-melted tracks were observed at low Volumetric Energy Density (VED) values, while porosity and cracking were more prevalent at high VED. The study reaffirmed that VED should not be considered a reliable dependent parameter due to its non-unique outcomes.

For **CuCrZr**, the optimized parameters were found to be LP = 320 W, SS = 655 mm/s, and HD = 0.10 mm. The study also revealed that VED could not be relied upon as a sole process parameter due to the scatter in density values for the same VED. Mechanical properties varied significantly with post-processing treatments; Direct Aging Heat Treatment (DAH) provided the best mechanical properties with an YS of 410 MPa and UTS of 438 MPa. However, Solution Treatment and Aging (STA) increased ductility, making it suitable for applications requiring higher elongation.

5.3 Future Scope

Future research could focus on refining the modified Taguchi approach for more complex multi-objective optimization scenarios involving additional aerospace-grade materials. The development of a more comprehensive model that integrates the effects of varying VED across different materials and their corresponding mechanical properties would be valuable. Furthermore, the study could extend to investigating the influence of microstructural evolution during LPBF and its impact on long-term material performance under cyclic loading conditions, which is crucial for aerospace applications. Advanced in-situ monitoring techniques could also be explored to provide real-time feedback during the LPBF process, enabling dynamic adjustments to process parameters for improved part quality and consistency.

List of Publications

1. Kurre Prasanth Kumar Reddy, Boggarapu Nageswara Rao, Nazeemudheen MN, Manwatkar SK, Murty SVSN. Selection of Optimal Process Parameters to Obtain Defect-Free Builds in IN718 Made by Laser Powder Bed Fusion. J Mater Eng Perform 2023:1–14. https://doi.org/10.1007/s11665-023-08677-9
2. Kurre Prasanth Kumar Reddy, Boggarapu Nageswara Rao, Narayana Murty SVS, "Multi-objective optimization to specify optimal selective laser melting process parameters for SS316 L powder", Multidiscipline Modeling in Materials and Structures 2023;20:59–80. https://doi.org/10.1108/MMMS-06-2023-0213.
3. Kurre Prasanth Kumar Reddy, S Ramesh Kumar, Boggarapu Nageswara Rao, "Multi-objective optimization for minimize cutting force and power consumption in corn stalk chopping processes", Heliyon 2024;10. https://doi.org/10.1016/j.heliyon.2024.e35373.
4. K. Prasanth Kumar Reddy, Shyam Prasad Kodali, N. Rajesh, T. Parameshwaran Pillai, Boggarapu Nageswara Rao, "Optimal Abrasive Jet Machining Parameters for Glass Fiber Reinforced Plastics", Journal of Polymer and Composites. 2024; 11(08):247-255.

www.ingramcontent.com/pod-product-compliance
Lightning Source LLC
LaVergne TN
LVHW010737061126
851857LV00014B/1521
* 9 7 9 8 2 4 0 8 2 6 9 4 8 *